BE AT YOUR BEST
WHEN IT REALLY COUNTS
Shows how to use your natural biological rhythms to lead a
healthier, more productive life

Also by
Susan Perry and Jim Dawson

Nightmare: Women and the Dalkon Shield

BE AT YOUR BEST WHEN IT REALLY COUNTS

Setting your body clocks for maximum energy and health

Susan Perry and Jim Dawson

THORSONS PUBLISHING GROUP

First UK edition published 1990

First published 1988 by Rawson Associates,
Macmillan Publishing Company, New York and Canada

British Library Cataloguing in Publication Data

Perry, Susan
Be at your best when it really counts.
1. Man. Biological rhythms
I. Title   II. Dawson, Jim
612'.022

ISBN 0-7225-2206-1

*Published by Thorsons Publishers Limited, Wellingborough,
Northamptonshire, NN8 2RQ, England*

Printed in Great Britain by Mackays of Chatham, Kent

1 3 5 7 9 10 8 6 4 2

for Joyce and Liz

To every thing there is a season, and a time to
    every purpose under the heaven.
A time to be born, and a time to die; a time to
    plant, and a time to pluck up that which is
    planted;
A time to kill, and a time to heal; a time to
    break down, and a time to build up;
A time to weep, and a time to laugh; a time to
    mourn, and a time to dance;
A time to cast away stones, and a time to gather
    stones together; a time to embrace, and a time
    to refrain from embracing;
A time to get, and a time to lose; a time to keep,
    and a time to cast away;
A time to rend, and a time to sew; a time to keep
    silence, and a time to speak;
A time to love, and a time to hate; a time of war,
    and a time of peace.

—ECCLESIASTES

Contents

Acknowledgments

Writing a book has a rhythm of its own, and many people helped us keep a steady beat in the writing of this one. First, we would like to thank the many scientists and researchers who gave us their time and expertise, especially Franz Halberg, Bob Sothern, William Hrushesky, Timothy Monk, Michael Smolensky, Mitchel Kling, Margaret Moline, David Sack, R. Curtis Graeber, Cynthia Hedricks, Sharon Golub, and Mark Mahowald. Our special thanks to the Melpomene Institute for sharing their research on women athletes and the menstrual cycle.

We would also like to thank our tireless researcher, Beth Atkinson Eames, who spent endless hours at the University of Minnesota medical library, and Billy Steve Clayton, who so patiently helped us design the book's graphics.

Finally, we would like to thank our American publisher, Eleanor Rawson, for her gentle guidance in shaping the book and our agent, Heide Lange, who, as always, was a constant source of support and enthusiasm.

One

The Times of Your Life

There is a time for all things.
—WILLIAM SHAKESPEARE
Comedy of Errors

Everything we do today is governed by the clock. We eat, sleep, work, play, and even make love according to the rigid schedule society has imposed on us. Most of us eat in the morning, again at noon, and then face our main meal in the evening. We work when we are told to, not necessarily when we most feel like it. Exercise, if we do it at all, is usually wedged between the demands of office and home. Lovemaking is reserved for the most socially convenient time — usually late in the evening. At weekends, we feel obliged to play hard, often staying up late and sleeping in. On Monday, we drag ourselves back to work, coffee in hand, trying to shake off the Monday morning blues.

Is this our natural schedule? No! In fact, it's an unnatural way to live!

For, although we structure our lives according to the harsh ticking of society's clocks, each of us has rhythms deep within us that are set to basic and ancient timekeepers. These are the natural rhythms we should try to live by.

What exactly are these rhythms? They are a complex system of internal pacemakers that regulate everything about us, from our body temperature to our moods and behaviors. They strongly influence when we will be happy and depressed, patient and irritable, careful and accident-prone. They also affect our susceptibility and resistance to illness and stress.

The scientific term for the study of these biological rhythms is *chronobiology.* All living organisms, from mollusks to men and women, exhibit these rhythms. Some rhythms are short and can be measured in minutes or seconds; others last hours, days, months, or even years. The peaking of body temperature, which occurs in most people every evening, is an example of a daily rhythm. The menstrual cycle is a monthly rhythm. The increase in sexual drive in the autumn is an example of a seasonal, or yearly, rhythm.

Understanding your internal pacemakers can help you organize your life so you can work *with* your natural rhythms rather than against them. It can enhance your performance at work and play.

For example, an understanding of biological rhythms can help you learn:

• When you should and shouldn't schedule important meetings and appointments
• When you should begin a diet
• How to avoid the Monday morning blues
• When you should exercise to minimize injuries and maximize performance
• When you should be most alert for symptoms of breast cancer and other diseases
• How to schedule your classes or those of your children to coincide with peak periods of memory retention and other mental skills
• How to minimize the effects of jet lag during a business or pleasure trip
• When your sexual drive is at its strongest
• When you should take or avoid taking IQ and other tests
• How to time taking medicine to do the most good with the least side effects
• When you are most likely to get pregnant
• When you are most prone to depression and moodiness
• How to time medical treatments or trips to the dentist so you will experience the least amount of pain

The list goes on and on. Biological rhythms are an integral part of our daily lives.

Your Rhythm Awareness: A Quiz

• Do you see yourself as either a night person or a day person?

• Is there a particular time of day when you feel you are most productive, no matter what you are doing?

• Do your energies seem to slump in the early afternoon?

• Do you feel sexier at a particular time of day? Month (for women)? Year?

• Do your moods change with the seasons?

• Is exercising easier for you at a particular time of day?

• Does food taste better to you as your day progresses?

If you answered yes to any of these questions, then you have an awareness of your biological rhythms—or, at least, the effect that these rhythms have on the way you feel.

THE BEAT OF A TYPICAL DAY

To understand better just how biological rhythms affect our daily lives, let's follow a mythical couple, Harry and Helen, through an average day.

It's Thursday morning. At 7:00 A.M., the alarm clock goes off next to Harry and Helen's bed. Helen already has been awake for an hour, but has stayed in bed listening to the morning news on the radio, as she does every morning. Harry, on the other hand, is sound asleep when the alarm rings and rouses only grudgingly from an active dream. He awakes with an erection, his fourth (unbeknownst to him) of the night. The erection is not the result of his dream, but of the ebb and flow of blood to his penis, which occurred in approximately ninety-minute cycles while he slept.

Both Helen and Harry are reluctant to leave the bed. Their body temperatures are only beginning to rise, which makes the bed seem extra cozy. And the hormones controlling their sexual drive are at their daily peak, creating an even greater temptation to remain between the covers.

But it is a workday for them and a school day for their two children. Harry and Helen get out of bed and start their morning routines. Although Helen is not a bundle of energy at this hour of the morning, she is much more alert and cheerful than Harry is. This is because Helen's daily biological rhythms are on a more advanced schedule than Harry's. The hormones that awaken her body and elevate her mood began their daily climb a good hour before those of Harry.

While Helen is downstairs preparing her family's breakfast, Harry stumbles into the bathroom to shave. This is both a good and a bad time of day for this activity—good because Harry's sensitivity to pain is low early in the morning, but bad because his hands are the most jittery in the morning. Fortunately, Harry manages to shave without cutting himself this morning. He does not notice that his beard growth was at an all-week low.

For breakfast, Harry and the children eat oatmeal with brown sugar and milk. Helen, who is on a diet to lose weight, settles for a glass of orange juice and a piece of toast with a dab of butter. She reminds Harry to pick up some groceries on his way home from work for the lasagna dinner she is planning to prepare that evening.

Helen would be wiser to eat a big meal at breakfast and a small one at dinner time. Calories consumed in the morning are less likely to turn to body fat than those eaten in the evening.

At the breakfast table, Harry and Helen's son, Horatio, announces that he has a spelling test at school that day. Fortunately, he studied for the test right after school the day before—the best time of day for learning something that must be recalled at a later date. Their daughter, Hilda, also has a big event that day—an after-school rounders game. Both she and the other children involved in the game should be playing their best at that hour, for everyone's motor skills peak in the late afternoon.

Unfortunately, Harry will not be as lucky. The timing of an important event in his day will not fall at the most opportune hour as far as his biological rhythms are concerned. He must make a big presentation to his boss at a meeting scheduled right after lunch. Harry's alertness and mental skills will be experiencing an early afternoon dip, which may make it more difficult for him to respond quickly and completely to questions from his boss.

After breakfast, each member of the family goes his or her separate way: the parents to work, the children to school. While Harry spends the morning putting the final touches on his presentation, Helen double-checks some figures for a report she is preparing. It is a good time of day for such work; her maths abilities are at their best.

During her lunch hour, Helen changes into running clothes and jogs three miles through a park near her office building. Although she finishes the run in twenty-one minutes—a good time for her— she probably would have recorded an even faster time if she had scheduled the run after work, when her athletic abilities, including endurance, will be at their daily peak.

By midafternoon, Helen is doing some background reading for her report and preparing an outline. Harry, meanwhile, is reflecting on what went wrong during his presentation and making plans to present the idea again to his boss at a later date. These are good choices for afternoon activities, because the mind's ability to reflect on old knowledge and absorb new information peaks during the afternoon. By 4:00 P.M., however, both Helen and Harry find themselves watching the clock. Their perception of time is slowing down, and the final hour of the workday seems to drag on interminably.

After work, Helen hurries to her daughter's rounders game and then home to prepare dinner for her family. Harry has arrived earlier with Horatio and has already made the sauce for the lasagna. The sauce smells particularly good at this time of day, for everyone's sense perceptions, including smell, are at an all-day high. Tolerance for noise, however, is at an all-day low, and the normal sibling squabbling between Horatio and Hilda seems particularly irritating to their parents.

After dinner, Helen and Harry sit down with an evening cocktail to listen to a special jazz concert on the radio. Because their senses are more acute in the evening, Helen and Harry will enjoy the music more now than at any other time of the day. In addition, the cocktails taste stronger at this hour, although they do not make Harry and Helen as tipsy as they would have earlier in the day.

By 10:00 P.M., Helen's body temperature has begun to drop and she feels drowsy. By 10:30 P.M., she is in bed, asleep. Harry, meanwhile, must wait another two hours for his temperature and other internal rhythms to prepare him for sleep. He stays up until 1:00 A.M., watching a late night talk show.

YOUR INNER RHYTHMS

The changes that occurred in Harry and Helen during their typical day—changes in alertness, sense perception, and mental ability— are the result of real and measurable biological changes going on within their bodies. All of us undergo such rhythmic changes all the time. Internally, our bodies are constantly in flux.

Take body temperature, for example. Every day your temperature rises and falls a degree or two with clocklike precision. Even if you stay in bed all day, your temperature will follow its regular daily pattern. That's why a one or two degree change in your temperature may or may not be an indication of illness—it all depends on *when* you are taking the reading.

Blood pressure behaves much the same way. It, too, has a regular up-and-down rhythm that can vary over the course of the day by as much as 20 percent. So, if you happen to have your blood pressure checked at a time of day when your cycle is at its peak, the reading you get may indicate a health problem that is not necessarily there, or, more importantly, miss a problem that does exist. (For more details about the health consequences of blood pressure rhythms, see chapter 7.)

The idea that our bodies are in constant flux goes against the medical training of most of today's doctors. Since the 1930s, doctors have been taught to believe the body has a relatively stable, or *homeostatic,* internal environment. Any fluctuations within the body were to be considered either random and meaningless or a sign of disease.

As early as the 1940s, however, some scientists were having difficulty accepting the homeostatic view of the body. One such scientist was Franz Halberg, a brilliant young European scientist working in the United States. In the late 1940s, Halberg made a remarkable discovery. He noticed that the number of white blood cells in laboratory mice was dramatically higher and lower at different times of the day. Halberg charted these highs and lows, and discovered that they formed a predictable daily cycle, which he called a circadian rhythm (from the Latin for "around a day").

With that experiment, the modern science of chronobiology—the study of biological rhythms in living creatures—was born. Halberg and his colleagues, and gradually dozens of other scientists around the world, began the meticulous process of finding and charting other rhythms in both mice and men. What they found was nothing short of amazing. For example, the time of day at which a person receives drugs or an X-ray for cancer treatment can literally mean the difference between life and death.

Still, it is taking a long time for many medical specialists, especially doctors, to accept this challenge to the old homeostatic viewpoint. But, as the evidence in support of chronobiology becomes more persuasive—helped in large part by computers, which are making both the detection and documentation of rhythms easier and more accurate—the scientific and medical communities are beginning to rethink their old-fashioned ideas about the human body.

Resistance still exists, however. Rarely do doctors take more than a single blood pressure reading during a surgery visit. Nor are all psychiatrists aware of the seasonal, or cyclic, nature of some forms of depression.

But chronobiology, considered an odd, minor science just a few years ago, is now being studied in major universities and medical centers around the world. There are chronobiologists working for the National Aeronautics and Space Administration (NASA), as well as for the National Institutes of Health and other government laboratories.

Chronobiology is becoming part of the mainstream of science, and it is changing our way of looking at life and time.

Biological Rhythms Are *Not* Biorhythms

Don't confuse the science of biological rhythms with biorhythms. The two are as unlike each other as astronomy and astrology.

Biorhythms were the brainchild of Wilhelm Fliess, a late nineteenth-century German doctor who was—for a time, at least—a friend of Sigmund Freud. Fliess believed that everyone is bisexual and has distinct male and female character traits. According to Fliess, the so-called male traits (such as strength, endurance, and courage) express themselves every twenty-three days, while the so-called female traits (sensitivity and intuition) have a slightly longer cycle of twenty-eight days. (Disciples of Fliess later added a third intellectual cycle of thirty-three days.)

By doing some mathematical calculations based on date of birth, Fliess believed everyone could predict his or her physical (male) and emotional (female) ups and downs. Fliess's theory might have been ignored if Freud had not at first proclaimed it to be a great breakthrough in biology. Eventually, however, Freud changed his mind and dismissed biorhythms as mathematical mumbo jumbo, but not before many other people had become true believers.

Because biorhythm charts are easy to put together, the validity of the biorhythm theory is relatively easy to disprove—and scientists have done so many times. For example, several years ago the US Federal Aviation Administration and the US National Transportation Safety Board examined the records of 8,625 pilots who had been involved in air mishaps. They found absolutely no correlation between negative biorhythms and the accidents.

It would be nice if determining our physical and emotional ups and downs were as simple as computing a few mathematical formulas based on our date of birth. But, the truth is, the rhythms within us are more complex than that.

THE NATURE OF RHYTHMS

If you think about it, the fact that your body has many internal rhythms shouldn't be surprising. Rhythms are all around us, and they have been since the beginning of time. Night follows day. The moon waxes and wanes in the night sky. The tides rise and fall. The seasons change.

The rhythms within your body are intricately connected with the cycles in the world around you, especially with the rising and setting of the sun. In fact, most chronobiologists believe that our internal rhythms originally evolved in response to the light and dark cycle of the sun. Having such rhythms enabled our earliest ancestors to anticipate changes in the physical world around them and, thus, protected and preserved the species. The peaking of our senses in the evening, for example, ensured that our ancestors would be extra alert to predators during the dangerous twilight hours.

At first, our internal rhythms were probably triggered by the sun. But gradually, over the course of evolution, the rhythms became deeply embedded within our bodies. And, although they have never lost touch with the sun's cycles, they can now run their courses separate from the influence of the sun.

One of the first scientists to discover this independent nature of biological rhythms was the eighteenth-century French astronomer Jean Jacques d'Ortous de Mairan. In 1729, he conducted a small experiment in which he observed that the leaves of a heliotrope plant, which normally open in the morning and close at night, continue that pattern even if the plant is kept in darkness for twenty-four hours.

De Mairan's experiment showed that something within the plant itself, not the sun, was responsible for the predictable opening and closing of its leaves. Scientists now believe that the mysterious something resides in the plant's genes. In other words, the rhythms of living things—plants and animals alike—are inherited. They are part of our genetic makeup.

BREAKING AWAY FROM THE SUN

De Mairan's work with the heliotrope plant was largely ignored for two hundred years. When scientists finally took it up again, they discovered yet another remarkable fact about biological rhythms: These rhythms have a daily cycle length of their own that rarely coincides with the exact twenty-four-hour cycle of the sun. In one type of plant, for example, the daily cycle may be 24.4 hours, while in another, the cycle may be 25.6 hours. Curiously, crossbreeding two plants with two different cycles produces a third plant with a cycle about halfway in between—a strong indication of the inherited nature of biological rhythms.

Scientists then started looking at the daily cycles of animals and found the same phenomenon. For example, a particular type of squirrel caged in constant darkness ran on an exercise wheel precisely every twenty-four hours and twenty-one minutes.

Thanks to the efforts of volunteers who have spent weeks (and sometimes months) in underground caves or windowless apartments, cut off from all outside time cues, we now know that human rhythms also will operate free of the sun's cycles. Most of us tend to have a natural daily cycle that is slightly longer than the twenty-four-hour cycle of the sun (although a rare few individuals have cycles shorter than twenty-four hours). In other words, our natural inclination is to go to sleep and wake up later each day. In this respect, we are like other animals that are active during the day; animals that are active at night, on the other hand, have a natural tendency toward a daily cycle that is shorter than twenty-four hours. No one knows why.

Also puzzling to scientists is the fact that women tend to have slightly shorter daily cycles than men, and—at least in the carefree world of an underground bunker—they also tend to spend a greater portion of their cycles sleeping. In addition, people who live alone tend to have shorter daily cycles than people with roommates; again, no one knows why.

Why have we broken away from the strict twenty-four-hour cycle of the sun? We may never know for sure, but some chrono-biologists speculate that having biological days that differ slightly from solar days makes us more adaptable to seasonal changes of light and darkness.

KEEPING IN SYNC

If our daily cycle is twenty-five rather than twenty-four hours long, how do we manage to fit our lives into the shorter schedule dictated by the sun?

Fortunately, our bodies are able to reset themselves each day to the twenty-four-hour rhythm, thanks to many powerful time cues. Chronobiologists call these cues *zeitgebers,* German for "time givers." Some can be found outside our bodies, some are located within, and others are part of our daily habits.

If our bodies did not respond to these time cues, society would be strange indeed. The concept of a day would mean something different to each of us. For some, a day might be 24.6 hours; for others, it might be 23.8 hours. For a few with weak body rhythms, it might even vary from night to night—say, 22.5 hours one day, 25.1 hours the next. Eventually, everyone's days would be out of sync with everyone else's. The result would be chaos. Setting up a meeting with a friend for next Saturday, for example, would be extremely difficult, for whose Saturday would you mean?

Zeitgebers, therefore, play a crucial role in our lives. The most obvious—and important—external zeitgeber is the sun's appearance and disappearance in the sky. Seeing the sun rise in the morning tells your body that it's time for certain internal rhythms to get underway—even though their natural inclination would be to wait another hour or so. Similarly, the approach of darkness is a cue for other rhythms either to rise or fall.

Other external zeitgebers provided by nature include such things as the chirping of birds in the morning and the rise and fall of the air temperature. Even electromagnetic fields may play a role in keeping our internal rhythms in sync. None of these, however, appears to be as strong a zeitgeber as the sun.

Outside time cues not only synchronize our bodies to the solar day; they also help our bodies anticipate the changing seasons. For example: Our bodies can tell by the length of night when winter is coming—and make the necessary adjustments in body rhythms, such as slowing down our metabolism so we keep more fat on our bodies as protection against cold weather.

For people living in the modern world, however, nature's time cues may be only slightly more important than those imposed by society and our daily habits. Setting an alarm clock, working a regular nine-to-five job, eating meals at the same time each day, even having a cup of coffee every morning—these activities help our bodies keep on a regular twenty-four-hour cycle.

As if we didn't have enough zeitgebers to keep our bodies in sync with the world, our internal rhythms also help synchronize each other, for none of the myriad rhythms within our bodies works in isolation. Some rhythms rise while others fall—like a modern dance in which the dancers move seemingly independently of each other, but which actually has been carefully choreo-graphed. The dance is so complex that chronobiologists are only beginning to understand the interrelationships of the rhythms.

THE DANGERS OF GETTING OUT OF SYNC

While we may not be consciously attuned to the complex time-keeping going on within us, we can become painfully aware of the system when it is thrown out of sync. The classic example of this is jet lag. When we travel across time zones, our bodies go crazy trying to adjust to the new time cues. Gradually, one by one, our body's rhythms reset themselves to the new cues; but it can take two weeks or more. In the meantime, we often feel downright awful, both physically and mentally.

Shift workers—people who work, say, 7:00 A.M. to 4:00 P.M. one week and then 10:00 P.M. to 7:00 A.M. the next—suffer from a kind of work-induced jet lag. Each time they change from one shift to another, their bodies have to reset themselves to new zeitgebers. As a result, their bodies never get to settle into a permanent rhythm. The same is true of people who work nights during the work week, but who then try to be day people on the weekends. Their rhythms are constantly reshuffling.

You don't have to travel across time zones or work the night shift to feel the effects of out-of-sync rhythms. Anyone who has suffered through the Monday morning blues has had an out-of-rhythm experience. Here's what happens: Over the weekend, you stay up late and sleep in late. By Sunday night your daily circadian rhythm has been pushed back a few hours. Falling asleep at a reasonably early hour is therefore difficult on Sunday night. When your alarm goes off on Monday morning, you feel like you're waking in the middle of the night. And, in a sense, it *is* the middle of the night—for your body's rhythms, at least.

Being out of rhythm is more than uncomfortable. It is also unhealthy. Numerous studies involving everything from plants to fruit flies to humans have shown that living things grow faster, produce more, and are simply healthier when their internal rhythms are in sync with their external environment. People who work changing shifts or frequently cross time zones (or do both, as airline pilots do) complain of a host of physical problems, including nausea, diarrhea, headaches, burning eyes, leg cramps, menstrual irregularities, and chronic sleep problems. They also suffer from more marital and emotional problems than the majority of people.

Keeping an irregular schedule may even shorten your life. In one study, mice that were subjected to shiftlike conditions had a life span that was 6 percent shorter than that of mice kept on a steady daily routine. In another experiment, a group of blowflies had their period of exposure to daylight shortened or lengthened once each week, just as if they had flown from New York to London. They died earlier than a second group of blowflies, who experienced a steady routine of twelve hours of sunlight followed by twelve hours of darkness.

Equally worrisome is the mental fatigue that occurs when rhythms are forced to reset themselves repeatedly. A study done by the United States Army, for example, found that soldiers flown to Europe needed as long as seven days to recover their ability to think clearly and logically. But most shift workers change shifts weekly. As a result, many suffer from continuous brain fatigue.

This kind of fatigue can, of course, lead to serious safety problems—for the public as well as for the individual workers involved. Several major plane crashes have been attributed to errors made by pilots who had been flying irregular hours before their final, fatal flights. The series of mistakes that led to the nuclear power plant disaster at Three Mile Island were made at 4:00 A.M. by people who were on weekly rotating shifts. The accident at the Chernobyl nuclear power plant in the Soviet Union apparently involved similar circumstances.

(For more information about what you can do to lessen the effects of jet lag and shift work, see chapters 8 and 9.)

How Flexible Are Your Rhythms?

Not all people find shift work, jet travel, or staying up all night discomforting. That's because some people have more flexible biological rhythms than others. In other words, some people have rhythms that adjust with relative ease to new time cues.

Unfortunately, flexible rhythms are not something you can acquire. You either are born with them, or you're not. Equally unfortunately, the flexibility of your rhythms diminishes as you grow older.

Although you can't make your rhythms more flexible, you can benefit from understanding just how flexible—or rigid—they are. If you have rigid rhythms, for example, you should probably stay away from shift work. On nights before an important meeting or other critical event, you should avoid staying up past your regular bedtime—even if it's by only an hour. Otherwise, your performance the following day will be less than your best.

On the other hand, if you discover you have flexible rhythms, you may not need to be quite so hard on yourself about keeping a strict daily routine. More than likely, your rhythms will adjust quickly to whatever (within some limits!) new zeitgebers you throw at them.

How can you determine your body's flexibility? Here are some questions to ask yourself:

• Do you prefer to do your work in the morning, rather than later in the day?

• If you stay out late at a party, do you find it difficult to sleep late the following morning, even if there's nothing to keep you from doing so?

• If you have little sleep one night, do you feel drowsy and sluggish the following day?

• Do you have a strong preference for eating your meals at fixed times?

• When you go away on vacation, do you stick to your normal routine for going to sleep and getting up?

• After several days of setting your alarm clock to wake you up at a certain time, do you find yourself waking up just *before* the alarm goes off?

• After flying across three or more time zones, do you find it difficult to adjust to the schedule of your new surroundings?

The more times you responded yes to these questions, the more rigid your rhythms. A string of no answers indicates that your rhythms are probably fairly flexible.

A VARIED BEAT

We've been talking mostly about daily rhythms. But your body has a variety of other beats, as well. Some are obvious, such as seasonal rhythms; others are more subtle, such as weekly rhythms. All play a part in your health and happiness.

Here's a closer look at the five major categories of rhythms.

Daily Rhythms and the Fountain of Youth

Because they are easy to detect and measure, more is known about daily—or circadian—rhythms than about any others. The most obvious circadian rhythm is the sleep/wake cycle. But there are other daily cycles as well: temperature, blood pressure, excretion of hormones, division of cells, and more. In fact, all functions of the body are believed to be governed by some kind of daily cycle.

Although each of us is born with a set of individualized circadian rhythms, it takes weeks—sometimes, years—for all the rhythms to become synchronized with the environment. One of the first rhythms to set a regular beat is urine flow. By their second or third week, infants wet their nappies more heavily during the day than they do at night—no matter how much breast or bottle milk they drink or when they drink it. Between one and five months of age, infants develop a daily heart rate rhythm. Between five and nine months, temperature and blood sugar rhythms can be clearly detected. But other body functions take longer to become established. For example, adrenal hormones—the alerting chemicals that stimulate your heart to pump harder, among other things—don't settle into a daily pattern until around the age of three, which may be why babies and toddlers are able to sleep easily during the daytime as well as at night.

Some chronobiologists believe that pediatricians will one day be able to judge how well a baby is developing by noting when the baby's various circadian rhythms become synchronized with the environment—just as pediatricians now look for such developmental milestones as walking and talking.

Our circadian rhythms also make some major changes at the other end of our life span. If you have a free-running circadian rhythm of 25.6 hours in your twenties, you could very likely have a free-running rhythm of 24.3 hours in your sixties. In other words, your daily cycle of sleeping and waking will most likely shorten as you age. In addition, the amplitude—the difference between the highs and lows—of many of your other daily rhythms, including body temperature, will shrink as you grow older. They may even disappear altogether.

Some chronobiologists believe that the secret to the fountain of youth may lie in keeping our circadian rhythms from shrinking and shortening as we grow older. So far, though, scientists are a long way from understanding just how that might be accomplished.

Seasonal Rhythms and Winter Madness

Some Eskimos suffer from a kind of annual "madness" known as Arctic Hysteria, which can last from a few hours to a few days every winter. Researchers who have studied this strange phenomenon discovered that the people afflicted with the "madness" secreted an unusually high amount of calcium in their urine during the winter months—about eight to ten times more than they secreted in the summer. Calcium strongly affects the nervous system; it is needed for the transmission of nervous messages throughout the body. It may be that the loss of so much calcium causes a temporary mental illness every winter.

Eskimos are not the only people who suffer from seasonal mental illnesses. Many people, particularly those in more northerly latitudes, suffer from an illness known as Seasonal Affective Disorder (SAD), a form of depression that strikes every autumn and winter as the days shorten and seems to be connected to another body chemical known as melatonin. In fact, some scientists believe that all of us experience at least a slight drop in our mood during the winter season. (For more about the relationship between biological rhythms and depression, see chapter 4.)

We have other seasonal—or *circannual*—rhythms, as well. Our sex drive, for example, peaks in late summer and early autumn—not spring, as poets would have us believe. Children grow faster in summer than in winter. Summer is also the time of year when our lungs and muscles are most efficient—and, therefore, when we are most likely to achieve a personal best at our chosen sport.

Even death seems to have seasonal preferences: deaths from arteriosclerosis peak around January; from suicide, around May; and from accidents, in late summer.

Just as circadian rhythms are tied to the earth's rotation on its axis, seasonal rhythms are connected to the earth's orbit around the sun. And just as sunlight (and the absence of it) helps synchronize our daily cycles, so, too, does it help synchronize our annual cycles. Through a process called *photoperiodism,* our bodies are able to measure the length of days and, thus, learn exactly where we are in the solar year. And *that* enables our rhythms to anticipate and prepare for changes in the environment.

In fact, seasonal rhythms can be easily fooled with high-intensity artificial lights. Farmers, for example, house chickens in window-less sheds and then use artificial light to trick the birds' hormonal systems into thinking it is spring, which causes the chickens to lay more eggs. Some psychiatrists have successfully used special high-intensity artificial lights to treat seasonal depression.

The Moon, Madness, and Monthly Rhythms

For centuries, all sorts of strange behavior has been attributed to the moon. Primitive people ascribed to the moon a host of mystical powers, from causing pregnancy to driving people crazy. In fact, the word *lunatic* comes from *luna*, Latin for moon.

Today, people continue to blame the moon—particularly, the full moon—for many things, from the birth of babies to murder and mayhem. It is not unusual, for example, to hear police officers talk about the extra craziness they run into on the streets when the moon is full. Some nurses who work in the maternity wards of hospitals claim that they know a full moon has returned by the dramatic increase in births, which they say happens then.

One successful New York stockbroker says that the moon even affects how people trade on the stock market. He says he has noticed that, when the moon is full, people get uneasy, restless, and tend to trade more. Usually, he adds, the trades they make then are wrong!

Whether or not there is any scientific truth to the folklore surrounding the moon remains to be seen. So far, studies looking into the effect of the moon on our behavior have been anything but conclusive. Some, for example, show a direct connection between the moon's phases and the incidence of homicide, suicide, and mental illness. Others show no such connection.

There is evidence, however, to support the idea of other monthly rhythms, ones deep within us that probably evolved millions of years ago in response to cyclic changes in the gravitational pull of the moon on the earth. That pull is dramatic. Not only does it cause our oceans to rise and fall, but it also causes the earth's crust to bulge and recede by as much as sixteen inches.

Our internal monthly rhythms are most likely synchronized to the pull of the moon and, to a lesser extent, to monthly changes in moonlight.

The most obvious of these monthly rhythms—which chronobiologists call *circatrigintan rhythms*—is the menstrual cycle of women. The average menstrual cycle is 29.5 days—the exact length of the lunar cycle. Because of the biological changes that are part of the reproductive cycle, women also experience a host of other monthly rhythms, from swings in mood and sexual desire to changes in susceptibility to illness. (For more about the ups and downs of the menstrual cycle, see chapter 5.)

Men may also have monthly rhythms. Unfortunately, few scientists have explored this possibility. One who has is Bob Sothern, a chronobiologist at the University of Minnesota who has kept meticulous track of his personal rhythms for more than two decades. Three to five times a day, every day, he stops what he's doing for ten minutes to measure the state of his body and mind. Using medical equipment he carries with him in a briefcase—a kind of portable doctor's office—he tests such things as body temperature, blood pressure, time perception, and hand-eye coordination.

Sothern says he has noticed several monthly cycles among his rhythms. The amount of air he can hold in his lungs and the strength with which he can grip an object, for example, appear to rise and fall monthly. So does the speed with which his beard grows.

Monthly rhythms may not be as pronounced in men as they are in women, he stresses, but they do seem to exist.

Mysterious Weekly Rhythms

Weekly rhythms—known in chronobiology as *circaseptan rhythms*—are one of the most puzzling and fascinating findings of chronobiology. Daily and seasonal cycles have an obvious link to the sun, and monthly cycles appear to be connected to the moon. But what is there in nature that would have caused weekly rhythms to evolve?

At first glance, it might seem that weekly rhythms developed in response to the seven-day week imposed by human culture thousands of years ago. However, this theory doesn't hold once you realize that plants, insects, and animals other than humans also have weekly cycles.

It may be that weekly rhythms simply evolved on their own. Living organisms may have found that they needed a cycle for their various internal processes that was longer than a day but shorter than a month. As a result, they "invented" the weekly cycle. Or, as chronobiologist Franz Halberg puts it: "We told Mother Nature we would rather do it ourselves." Mother Nature was not completely forgotten, however. The seven-day cycle is synchronized with both the sun and the moon. It is, after all, made up of twenty-four-hour cycles, and four seven-day cycles roughly equal one lunar month.

Biology, therefore, not culture, is probably at the source of our seven-day week. It certainly is a rhythm deeply ingrained within us. After the French Revolution at the end of the eighteenth century, the new rulers of France tried to abolish the seven-day week, which they thought was rooted in religious superstition, in favor of a more "rational" ten-day week. But the experiment failed. People continued to take a day of rest every seven rather than every ten days. More than one hundred years later, revolutionary leaders in the Soviet Union made a similar attempt to change the week—first to five days, then to six. Again, people resisted, and the seven-day week eventually was reestablished.

Scientists now theorize that our social week may actually be a zeitgeber, helping to reset our weekly biological rhythms—just as our daily social routines help reset our daily rhythms. That may explain why some of us who are used to relaxing on Saturday and Sunday feel so disoriented when we must work through the weekend. We have disrupted our rhythms.

Weekly biological rhythms are difficult to detect. In fact, chronobiologists were not even sure that such rhythms existed until fairly recently. But weekly rhythms have been found throughout the body, including such basic body functions as blood pressure, heartbeat, and oral temperature. Weekly patterns have also been discovered in the rise and fall of several body chemicals, including cortisol, the hormone that helps your body cope with stress.

In fact, weekly rhythms appear easiest to detect when the body is under stress, such as when it is defending itself against a virus, bacterium, or other harmful intruder. For example, cold symptoms (which are really signs of the body defending itself against the cold virus) last about a week. Chickenpox symptoms (a high fever and small red spots) usually appear almost exactly two weeks after exposure to the illness.

The body also seems astonishingly vulnerable at seven-day intervals. Long before penicillin was discovered, doctors knew that pneumonia and malaria patients were at greatest risk around the seventh day of their illnesses. Today, doctors who perform organ transplants know that their patients are in the most danger of rejecting their new organs at seven-day intervals following their operations.

Ultradian Rhythms and the Ninety-Minute Enigma

Rhythms in your body that occur more frequently than every twenty hours are known as *ultradian rhythms*. (They're called "ultra" because their frequencies are high.) Some ultradian rhythms are very short. Your brain waves, for example, have a fraction-of-a-second pulse, your heart beats in one-second intervals, and your breathing follows a six-second rhythm.

Others are longer. One of the most fascinating is the basic rest-and-activity cycle of about ninety minutes. When you're awake, your ability to concentrate wanes and your tendency to daydream increases about every ninety minutes. A similar pattern occurs when you sleep: You dream about every ninety minutes or so. There are other ninety-minute ultradian rhythms, as well. Hunger pangs, for example, come and go in ninety-minute cycles. So does the impulse, among many smokers, to light up a cigarette.

Interestingly, when you are under stress, these ninety-minute rhythms shorten, sometimes to sixty minutes or less. That is probably why we chain-smoke or eat more frequently during high-stress periods of our lives.

The reason for ninety-minute cycles is not clear, but it may be a way for nature to ensure that we have regular periods throughout the day when we are alert to our physical needs and the dangers around us.

Your Inner Rhythms: Some Examples

Type of Rhythm	Length	Examples
Ultradian	Less than twenty hours	Heartbeat 90-minute fluctuations in energy levels and attention span Brain waves
Circadian (daily)	About a day	Temperature Blood pressure Sleep/wake cycle Cell division
Circaseptan (weekly)	About a week	Rejection of kidney, heart, and pancreas transplants Blood pressure Heartbeat Common cold
Circatrigintan (monthly)	About a month	Menstrual cycle
Circannual (annual)	About a year	Seasonal depression Sexual drive Susceptibility to some diseases

FINDING YOUR RHYTHMS

Your body's rhythms affect every aspect of your life, from your moodiness to your sexual desire to your ability to diet or give up smoking.

As a matter of fact, it is even affecting your ability to read this book! For example, if you are reading this book in sittings of ninety minutes or longer, you may find your concentration periodically waning as you go in and out of the daydreaming stage of your ninety-minute ultradian cycle. If you are reading this book in the afternoon, you probably will remember more of the material presented here than you would if you read it earlier or later in the day.

In the chapters that follow, we'll look in more detail at the fascinating and little-known ways in which your body's internal rhythms affect your life. We'll start with your everyday rhythms—specifically, those that make up the wake half of your daily sleep/wake cycle.

Two

~

Your Daily Ups and Downs

Think in the morning. Act in the noon.
Eat in the evening. Sleep in the night.
—WILLIAM BLAKE,
The Marriage of Heaven and Hell

When Maggie and Steve got married, they seemed ideally suited
for each other. They came from identical backgrounds, held similar
values, and enjoyed many of the same interests and tastes, from
renovating old houses to preparing spicy ethnic meals.

But within a year of their marriage, Maggie and Steve began to
notice one major difference between them: They hardly ever got
up or went to bed together. Maggie was usually up and active by
6:30 A.M. By 8:00 A.M., she had taken her morning run, showered,
eaten, and read the morning paper. Steve, on the other hand,
would lumber out of bed a few minutes before 8:00 A.M. and
would come down to the breakfast table groggy, grumpy, and
asking for coffee. Not until 10:00 A.M., about an hour after he was
at work, would he begin to feel alert and ready for the day ahead.

At night, the situation was reversed. Maggie was the one whose
energies sagged early; usually she was sound asleep by 10:30 P.M.
Steve, however, would stay awake well past midnight, reading,
watching television, or just puttering around the house.

At first, this difference in daily schedules seemed amusing to
Maggie and Steve. They teased each other about it. But as their
marriage continued, it became a major source of conflict between
them. Maggie began to accuse Steve of being lazy and disorganized
because he spent so much of the morning in bed or bumbled
through the early hours of the day. She became especially irritated
at weekends, when he would stay up late at night and then sleep

until almost noon. Steve, on the other hand, couldn't understand why Maggie fell asleep so early in the evening. It bothered him that he didn't have a companion after 10:00 P.M.

Their different schedules also began to affect their sex life. Maggie was too tired at night; Steve was too sleepy in the morning.

The marriage soured, each one blaming the other.

Maggie and Steve's problem is not an uncommon one. Many of us live with another person—whether a spouse, child, parent, or lover—who may be on a completely different daily schedule than ours. Until fairly recently, behavioral scientists believed that such differences were the result of personal eccentricities or habits developed early in life. Scientists have come to understand that each of us has a distinct rhythm of sleep and waking that is programmed into our genes.

In other words, being a "morning" person or a "night" person is something you are born with, like your height or the color of your eyes. Unfortunately, there is little you can do to change it.

You can, however, make it a lot easier on yourself—and on the people around you—by learning how to recognize and understand your individual sleep/wake cycle and a few other basic daily rhythms. This chapter will show you how to do just that. Specifically, we'll look at some of the key rhythms that make up the wake half of your sleep/wake cycle. Then, in chapter 3, we'll take an equally close look at the rhythms that make up the dark, or sleep, side of your daily cycle.

Once you recognize these rhythms, you can begin to chart a daily course that is more in step with their important underlying beats. As a result, you should find yourself leading a healthier and more productive life.

THE BEST OF TIMES, THE WORST OF TIMES

You are a much different person at eleven o'clock in the morning than you are at eleven o'clock in the evening.

So different, in fact, that trying to perform a particular task at a "wrong" point in your daily cycle can have the same effect as trying to perform that task after becoming legally drunk or after getting only three hours of sleep the night before!

Are You a Morning or a Night Person?

Chances are, you already know instinctively whether you are a morning person (sometimes known as a "lark") or a night person (sometimes called an "owl"). If you aren't sure which you are, here are some questions to ask yourself:

- Do you wake up early and go to bed early?
- Do you generally rise from your bed feeling wide-eyed and raring to go?
- Do you feel that you do your best work early in the day?
- Do you find yourself waking up just before your alarm is scheduled to go off?

If you answered yes to these questions, then you are most likely a morning person.

- Do you wake up late and go to bed late?
- Do you wake up sleepy-eyed and sluggish?
- Do you generally suffer through the early morning hours and get your surge of energy and creativeness later in the day?
- Do you find it easy to sleep through the buzz or ring of an alarm clock?

If you answered yes to these questions, then you are most likely a night person.

The reason, of course, is that your body is not biologically constant. The hormones and other body chemicals that affect your ability to think and move have their daily ups and downs within your body.

How you feel, how well you do work assigned you, your level of alertness, your sensitivity to taste and smell, the degree with which you enjoy food or take pleasure in music—all are changing throughout the day.

Chronobiologists have been successful in determining some general time frames within the day when different tasks are most likely to be the easiest for us to accomplish. Here's what they have determined:

When You Are Most Alert

Most of us reach our peak of alertness around noon. During the work week, therefore, you might be wise to postpone lunch until after this period of peak alertness—and put it to use on a work project instead. At weekends, you may wish to schedule errands or projects around the house late in the morning, since at this time of day you will feel most energetic and eager to get things done—and less like procrastinating.

When You Are Least Alert

Not surprisingly, we are least alert during the early morning hours—specifically, between 3:00 A.M. and 6:00 A.M. Our ability to think clearly and react quickly is at its lowest point then. Sleepiness is not the only reason for this daily slump in alertness. Other internal rhythms that affect our alertness also bottom out a few hours before dawn.

If you must be active during this period of the day, try to avoid doing things that require you to make accurate mental judgments or quick, physical responses. This is *not* the time to work on your tax return (even if it is due tomorrow), nor is it the time to be driving on a motorway. In one US study single-vehicle lorry accidents, for example, increased *sixteenfold* between 4:00 A.M. and 6:00 A.M.— even though many of the drivers involved in the accident had had a full eight hours of sleep the night before.

The Early Afternoon Letdown

Soon after our alertness reaches its peak around noon, it takes a sudden drop. We feel tired and less able to concentrate; we may even feel an uncontrollable urge to take a nap. Many people blame this daily dip in energy on the food they eat at lunch. That's not the complete explanation. Studies have shown that no matter what—or whether—you eat or drink at lunch, you still will experience an early afternoon letdown. But eating a large lunch, particularly one high in carbohydrates, just exacerbates the problem, since carbohydrates calm us down.

Some scientists believe that this circadian rhythm is linked to our tropical origins. Evolution probably built it into our bodies to protect us from overexerting, and thus harming, ourselves during the lasting heat of the day. Indeed, many warm-climate cultures throughout the world continue to shut down their offices, stores, and schools for a daily siesta during the early afternoon hours.

Fortunately, the early afternoon letdown lasts for only a couple of hours, at most. One way of lessening its impact on your work is to eat your lunch late—say, at 1:00 P.M. or later, and to include some energizing protein in it. Then, when you return to your job, try to schedule easy tasks until your alertness levels begin to rise again—usually by 3:00 P.M. Make sure those tasks aren't boring ones, however, or you may find yourself falling asleep on the job!

One final word about the early afternoon letdown: If you are a morning person, you may feel this daily sag in energy more than a night person. Scientists are still trying to determine why.

When Your Memory Is at Its Best

Your *immediate,* or *short-term,* memory is best during the morning hours—in fact, about 15 percent more efficient than at any other time of day. So, students, take heed: When taking an exam in the morning, it really does pay to review your notes *right before the test is given.* This has applications in the work world as well. You will need to rely less on written notes during a morning meeting than one held in the afternoon—if you review your notes right before the morning meeting. If it's important that you look like you can rattle figures or facts off the top of your head, you should try to do the rattling early in the day. Just make sure you review whatever it is you want to remember right before you go into the meeting.

Your *long-term* memory is a different matter, however. Afternoon is the best time for learning material that you want to recall days, weeks, or even months later. If you are a student, therefore, you would do best to schedule your important—or more difficult—classes in the afternoon, rather than in the morning. You also should try to do most of your studying in the afternoon, rather than late at night. Interestingly, many students are fooled into thinking that they study best while burning the midnight oil, because their short-term recall is better during the wee hours of the morning than it is in the afternoon. But short-term memory won't help you much twelve days or even twelve hours later, when you are facing the blank page of an exam book!

Of course, students are not the only ones who can benefit from scheduling their study periods in the afternoon. Lawyers, for example, would do best to do their trial preparation in the afternoon to ensure good recall of the material later in the courtroom. Politicians, business executives, or others who must learn speeches would do best to do their memorizing in the afternoon.

Businesses also can make the most out of this circadian rhythm. If you want to ensure, for example, that a new trainee will remember most of what is presented to him or her during a training session, offer the training session in the afternoon. This goes for coaching sports teams as well. If you want players to remember complex plays and instructions, then schedule team meetings in the afternoon.

Note: How well you remember things depends on when you *learn* them, not when you *recall* them. In other words, when you take a spelling test is not as important as when you studied for it.

When Your Thinking Is at Its Best

On average, we tend to do best on cognitive tasks—things that require the greatest mental effort—during the morning hours, particularly the late morning hours. Therefore, you would do well to schedule all of your work that requires complex thinking or organizational skills at that time of day. This includes, of course, a host of activities, from writing a report to designing a building to balancing your checkbook.

Because cognitive skills are high late in the morning, this is also a good time to schedule creative sessions or "working" meetings in which important ideas are discussed and evaluated. Just make sure the meeting breaks up before the early afternoon letdown sets in and everyone's energy and ideas begin to droop!

If you are a morning person, you also should be aware that during the late afternoon and early evening hours your reasoning skills will experience a considerable sag. This is definitely not a good time for you to take on tasks that require considerable complex thinking. Evening types also have a sag at this time of day; however, it is not as pronounced.

When You Are Best at Simple, Repetitive Tasks

If all or part of your job requires simple, repetitive tasks, you'll find that you probably do your best work around midafternoon. These are the tasks that are not too complex or that don't require you to draw too much on your memory—such things as filing, sorting, photocopying, and proofreading (but not conceptual editing, which requires more complex thought processes). Around the house, it might include such things as folding laundry, doing dishes, raking leaves, or other household tasks.

The tasks must be simple and straightforward, however, to fall into this category. Once the task gets even the slightest bit complex, it falls into the cognitive category and is best reserved for late morning.

When You Work Best with Your Hands

Your manual dexterity—the speed and coordination with which you can perform complicated tasks with your hands—is definitely best during the afternoon hours. Such work as carpentry, typing, sewing, or anything else that requires skilled use of your hands, will be easier at that time of day.

You might also find it a good time to practice the piano or other instrument—particularly since long-term memory skills are also high in the afternoon, which will help you in your memorization of a piece of music.

When Your Mood Is at Its Best—Maybe

Surprisingly, our moods—whether we feel happy or sad, calm or tense, patient or irritable—do not appear to have strong daily cycles. It's not clear why this is true. It may be because our emotions are easily affected throughout the day by the things we see and hear—for example, a particularly sad article in the morning newspaper—and that these outside factors override any inner rhythms. Our happiness quotient does appear to peak during the late morning hours, but chronobiologists are not sure whether this is a true rhythm or just part of the general well-being we feel at that hour because of our increased alertness.

This is not to say, however, that mood has no role in our daily rhythms. A bad mood (feelings of sadness, irritation, and tenseness) can sometimes be an indication that something has gone wrong with your cycles, particularly, that they have gotten out of sync with each other. That's why shift workers, for example, often feel so tense and irritable after joining a new shift: The link between their temperature and sleep/wake cycles has been broken. (For more about shift work, see chapter 9.)

If you begin to feel irritable or depressed for what seems like no reason at all, you may want to take a hard look at your daily routine. Is there anything in it you have changed recently, any reshuffling of zeitgebers? If so, you can either go back to your old routine or wait for your inner rhythms to catch up with your new one.

When Your Senses Are Most Acute

All of your *senses*—taste, sight, hearing, touch, and smell—are their keenest during the twilight hours of late afternoon and early evening. This is why dinner usually tastes better to us than breakfast, why bright lights irritate us at night, and why noisy children can drive us up the wall when we come home from work in the evening.

It may also be why societies throughout the world gather in the evening to talk, eat, and play music. This is the time when our rhythms are most attuned to sensory activities.

When You Are the Best Judge of Time

Not only does time seem to fly by when you're having fun, but it also seems to fly by even faster if you are having that fun early in the morning or late at night. That's because our perception of time changes during the day. The change is tied directly to the daily rise and fall of our body temperature. When our temperature is low (early morning and late evening), our perception of time speeds up; when our temperature is high (throughout the afternoon and into early evening), our perception of time slows down. (This is also why time seems to drag when you are ill with a fever.) Thus, it's not your imagination that it seems to take forever for the end of the working day to arrive. Well, actually, it *is* your imagination—but now you know that there's a reason for it!

When You Are Best at Sports and Other Athletic Activities

The best time to play tennis, ski, run, throw a baseball, or pursue any other sporting interest is during the afternoon and early evening. During this part of the day, your body is primed for optimal physical performance. It's when you are most coordinated, and when you're able to react the quickest to an outside stimulus— like a baseball speeding toward you at home plate!

Studies have shown that late in the day you will also *perceive* a physical workout to be easier and less fatiguing—whether it actually is or not. That means you are more likely to work harder during a late afternoon or early evening workout, and therefore benefit more from it.

The worst time for engaging in physical activities is, not surprisingly, between 3:00 A.M. and 6:00 A.M., when you are least alert, both physically and mentally. The next few hours are not much better. So, if you run or do some other physical workout early in the day, you may want to switch it to a later hour, preferably one during late afternoon or evening. You should see a notable improvement in your performance. Studies involving swimmers, runners, shot-putters, and rowing crews have shown *overwhelmingly* that performance is better in the evening than in the morning.

The exception might be for sports activities that require extreme exertion or must be undertaken under adverse weather conditions—such as participating in a triathalon on a miserably cold and wet day. We tend to tolerate physical discomfort more during late morning than at any other time of day.

Researchers at NASA and elsewhere have been able to pinpoint even smaller "windows" of physical performance during the day that can be applied to particular sports. Just when these windows occur depends on the kind of skill the sport requires:

For example:

• *Both the strength with which we can grip an object and the strength with which we can pull with our arms peak between 2:00 P.M. and 8:30 P.M.* This time frame, therefore, is good for weight lifting, rowing, tennis, racquetball, and other sports that require strength in the hands and arms.

• *Body flexibility peaks around 1:30 P.M.* Thus, divers, gymnasts, and other athletes who bend and twist their bodies should find it easier to perform during the early afternoon hours. Yoga practitioners, also take note!

• *Our hands tend to tremble more in the morning, especially around 11:00 A.M.* This is probably because our brain's alertness chemicals, dopamine and norepinephrine, are at their peak then. Perfection in sports that require fine neuromuscular control, such as archery or darts, can therefore be more elusive at this time of day.

• *Our aerobic ability is at its peak in late afternoon and early evening.* At that time of the day, our lungs and heart work more

How One Person Uses Her Rhythms in Her Work

Jackie is a free-lance writer who works out of her home. Because she is self-employed and essentially her own boss, she can set her daily work schedule. After some trial and error, she settled on the following schedule:

9:00 A.M. to 12:30 P.M.: Writing
12:30 P.M. to 1:00 P.M.: Lunch
1:00 P.M. to 3:00 P.M.: Interviewing, meetings, research
3:00 P.M. to 5:00 P.M.: Billing, filing, proofreading, correspondence

Jackie finds this routine to be the most productive one for her. She says that her writing seems to go better in the morning—the time of day she feels most alert and able to concentrate. "By midafternoon, my attention span begins to waver," she notes. "So I find it best to do less intense and shorter projects then, like writing letters to my editors or proofreading articles."

Although Jackie doesn't know it, the schedule she has set up for herself is attuned to her natural circadian rhythms. Most of us, as noted earlier, are better able to handle cognitive tasks, such as writing, during the morning hours. Repetitive tasks, such as filing and proofreading, are best done in the afternoon.

Jackie isn't the only one who has unconsciously scheduled her workday to the beat of her inner rhythms. A study of a group of English secretaries found that they, too, instinctively postponed their most repetitive tasks—filing and photocopying—until the afternoon hours.

efficiently, and our bodies are better able to draw upon their stores of energy. This time of day, therefore, is best for running, cycling, and other sports that require aerobic endurance.

Of course, most sports require a combination of these skills. Tennis, for example requires hand strength, trunk flexibility, *and* aerobic stamina. So, before you schedule an important grudge match, you may want to do some experimenting on your own to discover the time of day when you are at your best.

DEFINING YOUR DAY

Although all of us appear to follow the same general daily pattern of ups and downs in our level of alertness and our ability to perform certain tasks, the exact timing of those ups and downs varies from person to person. You may reach your daily peak of alertness around 11:00 A.M., but your spouse (or boss, or friend, or anyone else) might not peak until 12:00 noon. Or you may reach a peak in long-term memory retention at 3:00 P.M. while someone else may reach it at 4:00 P.M.

It all depends on how your *biological* day is structured—in other words, it depends on the timing of the circadian rhythms within you. To a great extent, that means how much of a morning person or how much of a night person you are.

In general, the earlier your biological day gets going, the earlier you are likely to enter—and exit—the peak times for performing various tasks.

Usually, the difference in the timing of a rhythm is relatively slight between a morning and a night person. Even an *extreme* morning person and an *extreme* night person (and only 20 percent of us fall into one of those categories) have circadian cycles that are no more than two hours apart. In fact, one study has revealed that extreme morning people go to bed at night and arise in the morning only eighty minutes earlier than extreme night people. Another study has shown that the body temperature of both groups tends to peak no more than sixty-five minutes apart.

When it comes to our reasoning powers, however, the differences between morning people and night people are much greater. Although both groups are at their best during the late morning hours, they can be miles apart at other times of the day. In general, a morning person should schedule his or her heaviest load of complicated, creative work throughout the morning—especially late morning—and avoid such work in the late afternoon. A night person, on the other hand, should avoid complicated reasoning tasks until late morning, and then not worry too much about a late afternoon sag.

Big differences between morning people and night people have also been found in the area of athletic performance. Although both groups tend to do best at sports between noon and 9:00 P.M., night people generally peak later during that time frame than morning people—in fact, by as much as five hours!

Differences Between Night and Day (People)

• Morning people tend to have more introverted personalities, while night people tend to be more extroverted. This is particularly true after the age of forty.

• Morning people tend to have less flexible circadian rhythms, which means they benefit more, both physically and mentally, from following a structured daily routine.

• Morning people tend to sleep more soundly than night people and wake up feeling more refreshed.

• Women are more likely to be morning people than men.

HOW TO DETERMINE YOUR "NIGHTNESS" OR "MORNINGNESS"

How can you find out just how much of a morning or night person you are? There are two tests you can use: One charts your alertness levels during the day, the other your temperature.

Tracking Your Alertness Rhythm

To find your alertness rhythm, simply take a minute or two every hour during your waking day to rate how alert you feel at that point. Use the five-point alertness scale shown here and record each reading in the spaces provided.

Continue the test for a minimum of three days. Then add up the hourly totals and divide by the number of days of the test to determine your average hourly reading. Plot that average reading on the chart below. This should give you a picture of when during the day you are most and least alert.

How to read the chart. Look for the time when your alertness peaks. Morning people tend to peak before noon; night people peak after noon. So the earlier your alertness peaks in the morning, the more of a morning person you are. The later your alertness peaks in the afternoon, the more of a night person you are.

Tracking Your Daily Temperature Cycle

The best way to determine your daily temperature cycle is to take your temperature every hour for twenty-four hours and then chart the results. Most of us, however, would find it impractical to take our temperature that many times during the day—not to mention waking ourselves up half a dozen times during the night! Fortunately, chronobiologists have developed a simpler method to help you figure out your daily temperature cycle. All you need to do is take your temperature five times during a single day and then

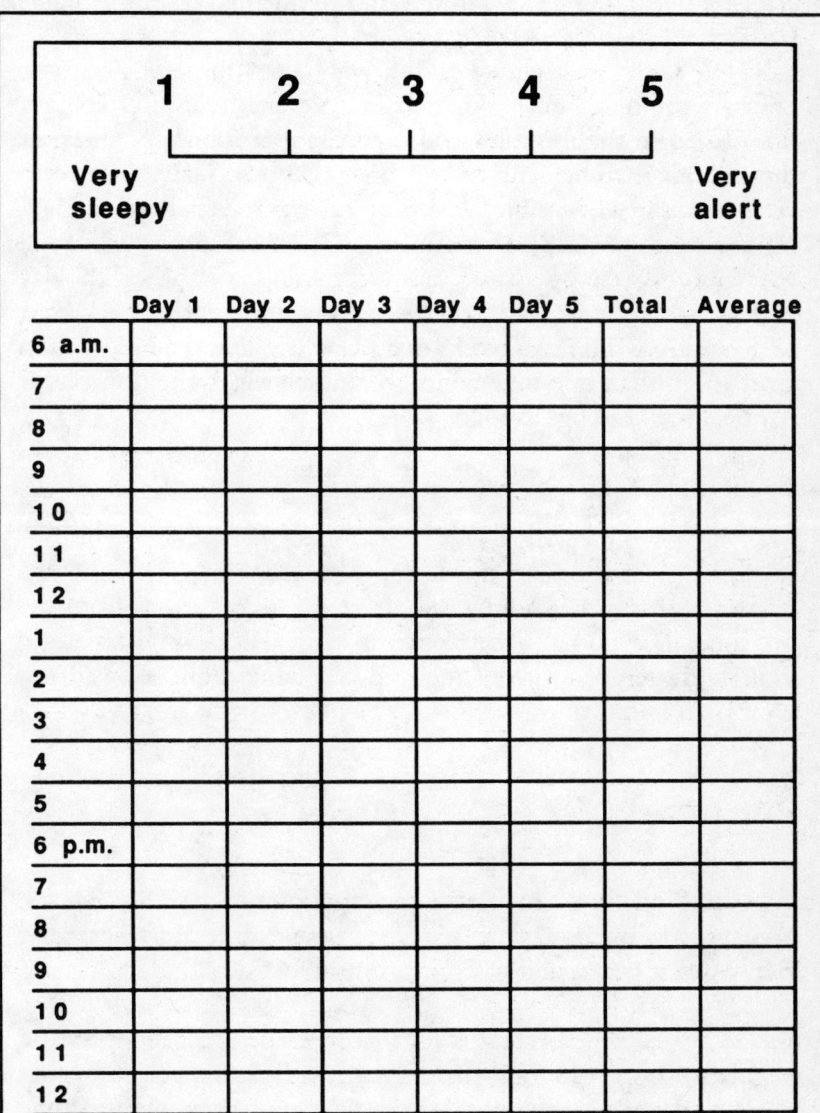

	1	2	3	4	5	
Very sleepy						**Very alert**

	Day 1	Day 2	Day 3	Day 4	Day 5	Total	Average
6 a.m.							
7							
8							
9							
1 0							
1 1							
1 2							
1							
2							
3							
4							
5							
6 p.m.							
7							
8							
9							
1 0							
1 1							
1 2							

estimate what your temperature is during the middle of the night. Here's how it works:

Step 1: Using an accurate oral thermometer (the new electronic ones are the best), take your temperature one hour after you get out of bed in the morning and then again at four-hour intervals throughout the day. The last reading should be taken as close to your bedtime as possible. A variety of drugs, including alcohol, caffeine, and over-the-counter medications, can make your natural temperature cycle go askew, so avoid these if possible on the day you are taking your readings. Also, do not attempt to track your temperature within two weeks of a plane trip that required you to cross more than one time zone. In a notebook, write down your temperature readings as follows:

Temperature Readings

1_____   2_____   3_____   4_____   5_____

Step 2: Add the first, third, and fifth readings. Then add the second and fourth readings. Subtract the second total from the first one. The number you come up with will be an estimate of your body temperature in the middle of the night. Consider it your sixth temperature reading.

$$_____ \quad (1, 3, \text{and } 5)$$
$$- _____ \quad (2 \text{ and } 4)$$
$$= _____ \quad (6)$$

Step 3: Plot all six temperature readings on the chart on page 43. You should now be able to see when your temperature begins to rise, when it peaks, and when it begins to drop.

How to read the chart. Look for the time when your temperature begins to drop. Morning people tend to have temperatures that begin to drop before 8:00 P.M.; the temperatures of night people usually drop after 8:00 P.M. The earlier your temperature drops in the evening, the more of a morning person you are; the later it drops, the more of a night person you are.

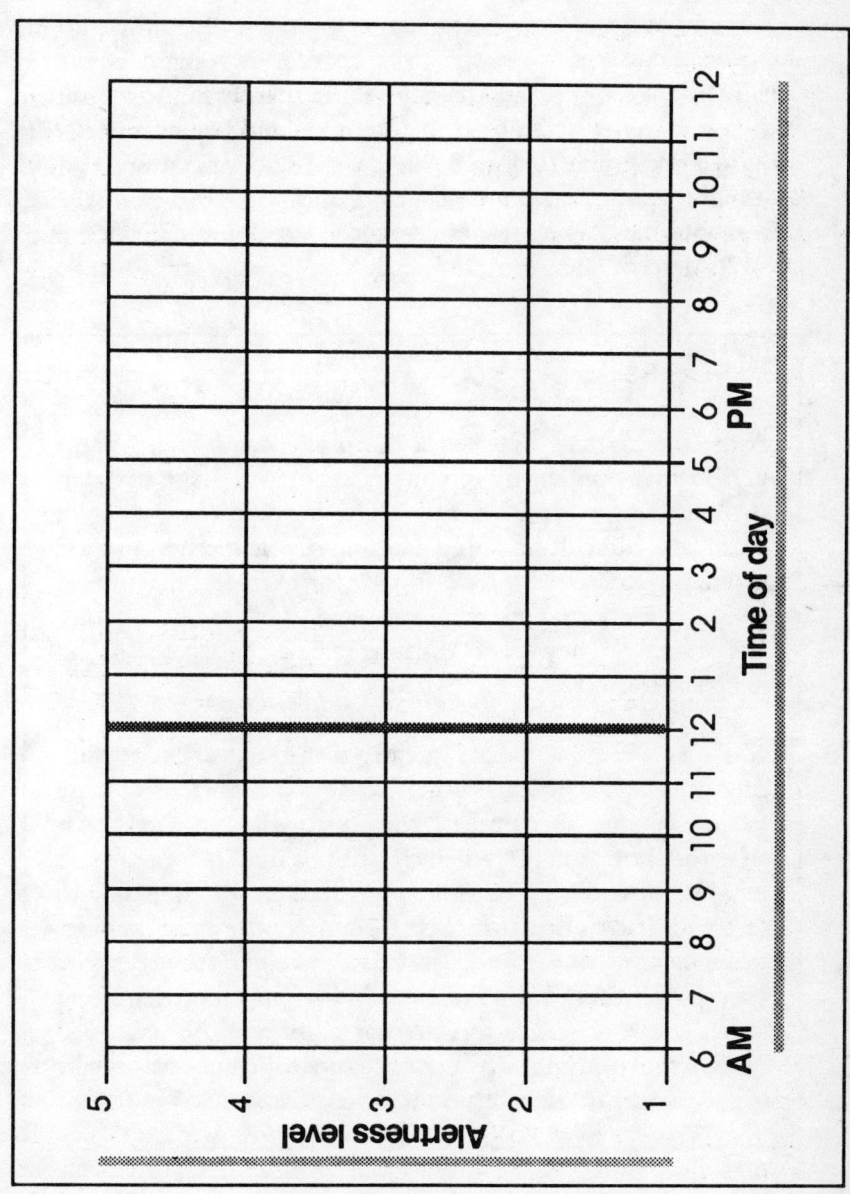

A tendency toward morningness or nightness also shows up in the general curve of the temperature cycle. A morning person will usually have a temperature that rises fairly sharply in the morning. It then reaches a plateau by early afternoon and begins its descent early in the evening (before 8:00 P.M.). The temperature rhythm of a night person, on the other hand, shows a more gradual rise throughout the day. It reaches its plateau late in the afternoon and starts its descent later, too (after 8:00 P.M.).

When a Lark Marries an Owl

A mixed marriage between a morning person and a night person may contain more conflict than a marriage between two morning people or two night people. These couples often have difficulty creating a feeling of togetherness in their marriage. Researchers have found, for example, that mixed couples have sexual relations less often, they go out together less often, and they spend less time talking to each other than matched couples.

If you and your spouse are one of these mixed couples, you may need to make a greater-than-usual effort to find time to spend with each other—time when both of you are feeling at your best. Embark on your nights out together at an earlier hour so you can enjoy a full evening before the lark member of your duo becomes too tired. Or set aside time to talk with each other during the afternoon or early evening, when neither one of you should be feeling too sleepy or grumpy. If you're having trouble finding time for lovemaking because one of you is too tired at night and the other too tired in the morning, why not try to get together occasionally in the middle of the afternoon? Be creative.

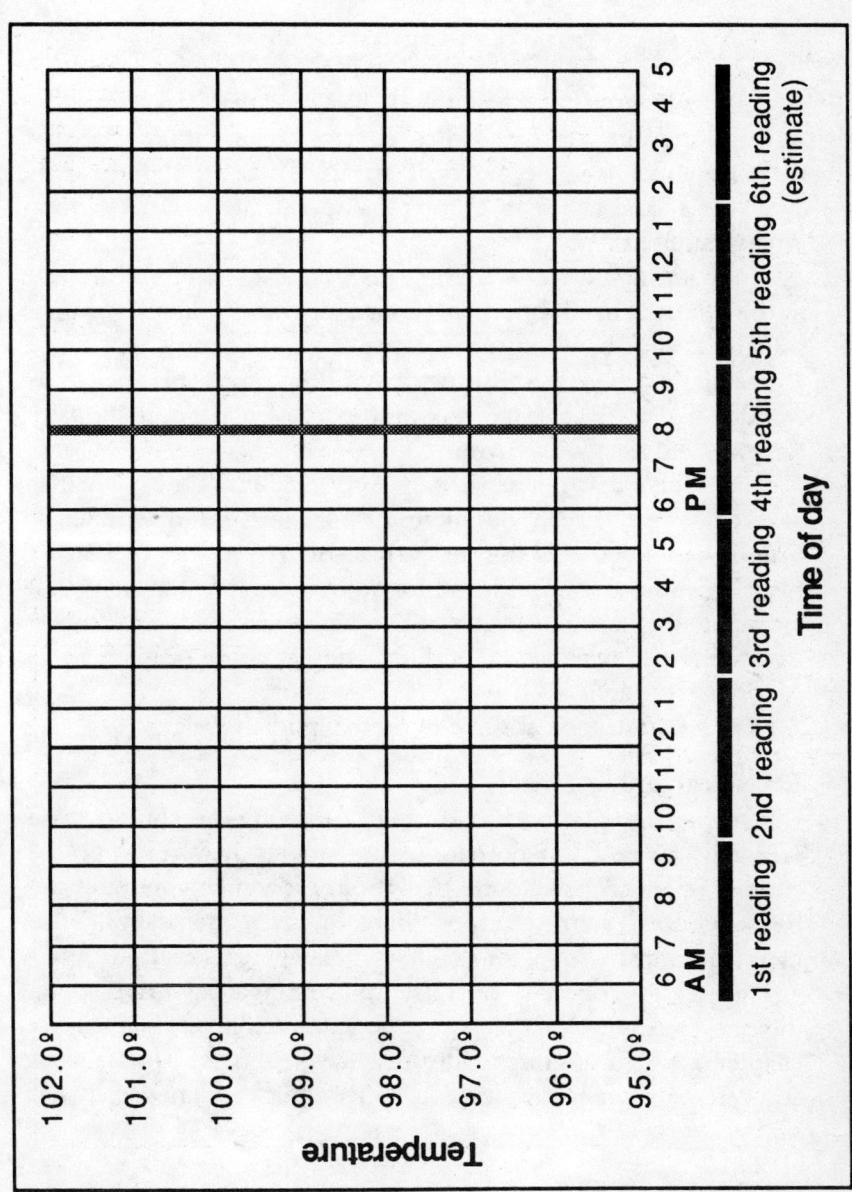

Temperature

102.0º
101.0º
100.0º
99.0º
98.0º
97.0º
96.0º
95.0º

6 7 8 9 10 11 12 1 2 3 4 5 6 7 8 9 10 11 12 1 2 3 4 5
AM   PM

1st reading 2nd reading 3rd reading 4th reading 5th reading 6th reading (estimate)

Time of day

YOUR NINETY-MINUTE CYCLES

So far we've looked only at the daily cycles that make up your waking day. But you have other, shorter cycles within your day that also affect your alertness and habits. These are the ultradian cycles we talked about in chapter 1—specifically, the mysterious ninety-minute rhythms.

Ninety-minute cycles were first discovered by sleep researchers in the 1950s. They found that about every ninety minutes, adults shift back and forth between deep sleep and dream sleep. Since our minds are more alert during dream sleep, researchers began to wonder if this sleep cycle was only part of a continuous daily alertness and sleepiness rhythm.

They found that it was. About every ninety minutes, whether we are awake or asleep, our alertness goes up—and down. Scientists have dubbed this our Basic Rest-Activity Cycle, or BRAC. More recently, researchers have found other ninety-minute cycles. However, they are not necessarily synchronized with each other. In other words, when one is peaking, another might be in midcycle or even at its nadir.

Here's a closer look at some of these ultradian cycles.

Daydreaming

When your mind wanders during a meeting or concert, you probably blame it on boredom. But biology may really be the culprit. Scientists have found that about every ninety minutes our bodies undergo subtle changes—in brain wave patterns, eye movement, and muscle tone, for example—that cause us to daydream.

These cycles appear to be a daytime continuance of the ninety-minute dream cycles that we experience while we sleep. (See chapter 3.) Like nighttime dreams, daydreams are spontaneous and often explore earlier life experiences, but they are usually less bizarre.

Sleepiness

The clearest ninety-minute rhythm is the "sleepability" cycle. About every ninety minutes, you enter a short period during which you are vulnerable to fatigue and sleepiness. If you were to lean back and close your eyes during one of these periods, you would find it easy to fall asleep.

Like the daydreaming rhythm, this susceptibility to sleep also seems to coincide with the nightly sleep cycle. It is the waking equivalent of deep sleep, the part of our sleep cycle when our minds are least active. (See chapter 3.)

These rhythmic periods of sleepability are shorter during the morning hours than in the afternoon. That's why it's usually more difficult to take a nap in the morning than later in the day. Some of the time, you can fight off this fatigue—particularly if what you are doing is interesting or different. However, most of the time you would do best to give in to the rhythm and take a short break from your work—especially around the lunch hour, when you are already battling the early afternoon letdown. Do a few stretching exercises or go for a short walk, even if it's only to the bathroom or water cooler. When you come back, the worst of the feelings of fatigue will probably have passed and your mind will be ready to concentrate once again on the task at hand.

It's also a good idea to keep this fatigue cycle in mind when you are traveling long distance by car. Be sure to stop and take a break every ninety minutes or so.

Going to the Bathroom

We tend to have a peak in urine flow about every ninety minutes. While that doesn't necessarily mean you will be rushing to the bathroom every hour and a half, it is something to keep in mind before going into a long meeting or movie. People who are planning conferences, take note: Make sure you have a break scheduled for your participants about every ninety minutes.

The Urge to Eat

If you keep track during the day of when you feel an overpowering urge to put food into your mouth, you'll find a regular ninety-minute pattern. Indeed, studies have shown that this rhythmic desire to eat is definitely not just in our minds; our stomachs contract (signaling us to eat) about every ninety minutes. There is a physical reason for munchie madness!

Of course, if you're trying to lose weight, this periodic urge to eat can undermine your efforts to stay away from food. You can help yourself by being aware of the cycle. Try to wait out the urge, or give in to it in a low-calorie way by munching on a carrot or cracker or by drinking water or a low-calorie juice. Remember, the urge will pass soon—fortunately.

(For more information about how to track your ninety-minute hunger pangs, see chapter 7.)

The Urge to Smoke

In addition to a ninety-minute hunger cycle, we also have a need every ninety minutes for oral satisfaction, an urge to put something into our mouths, such as a toothpick, cigarette, or stick of gum. This oral urge may or may not occur at the same time as the hunger cycle. Often, however, people satisfy this urge with food.

People who smoke tend to turn to cigarettes instead of food when the oral urge hits. This may be one of the reasons smokers often put on weight when they give up cigarettes: They now give in to the cycle by eating instead of by smoking.

Again, if you are trying to quit smoking, just being aware of this ninety-minute cycle of oral activity can help you resist the temptation to light up a cigarette. Indeed, several stop-smoking programs place great importance on closely charting the strength of your craving for each cigarette you smoke. Within a day or two, the cyclic nature of the habit becomes apparent.

It is important to anticipate these peaks and work through them. You can accomplish this with a variety of tricks, such as substituting a carrot or stick of chewing gum for a cigarette or doing deep breathing exercises. To do the breathing, get into as comfortable a position as possible (lying down is ideal). Close your eyes and mouth, then inhale slowly and deeply through your nose. Feel your lungs fill with air. Then, exhale slowly. Continue to do the deep breathing exercises until you feel your body relaxing and the urge to smoke abates.

HOW STRESS SHORTENS YOUR CYCLES

Scientists have found that when we are bored, under stress, or short of sleep, our ninety-minute cycles tend to shorten to roughly sixty minutes. That may explain why we eat and smoke more under stressful or boring circumstances.

Tips for Keeping Your Daily Rhythms on Beat

The more you keep your inner rhythms in sync with each other and with the environment around you, the better you will feel and perform. Here are some tips on how to do just that:

• Get out in the sun for at least fifteen minutes each day. Direct sunlight provides the strongest time cue for your rhythms and helps reset them each day.

• Try to follow a regular daily routine. This is especially important if you are a morning person, for your rhythms have more difficulty adjusting to change than those of a night person.

• Evaluate your work schedule. If you are an extreme night person, for example, who must report to work daily at 8:00 A.M., perhaps you should be looking for a new job—one in which you can work a later shift and, thus, be more in step with your natural rhythms.

• Find out the basic structure of your circadian day (see alertness and temperature charts on pages 41 and 43), and then try to schedule your daily activities to coincide with your peak periods of performance.

• Be aware of your ninety-minute cycles and use that awareness to avoid binging on food and to quit smoking.

Chronobiologists theorize that these shortened cycles are a regression toward the cycles of our infancy, since, as babies, our BRAC cycles are only about sixty minutes long. It also helps explain why babies demand to be fed so often!

So, when you are under stress, you may need to take even more breaks than usual from activities that require intense concentration. Also, be extra alert to the impulse to smoke or eat.

Three

~

The Importance of Sleep

We are such stuff as dreams are made on,
And our little life is rounded with a sleep.
—WILLIAM SHAKESPEARE
The Tempest

Young, clean-shaven Rip Van Winkle, so the story goes, fell asleep one evening in the Catskill Mountains only to awake bearded, befuddled, and bent with age twenty years later. While sleeping for that length of time may seem like a fantasy, it is actually closer to the truth than most of us realize. For we spend a full third of our lives—or about twenty-five years—on the dark side of our daily cycle.

Although our daily need to sleep is a rhythm we all recognize, sleep itself remains a great mystery. Even the most fundamental question about sleep—why we need it—can't be answered with certainty.

Yet sleep, along with food and water, remains one of the basic necessities of life. Without it, we become progressively irritable and depressed, we lose our powers of concentration, and our work and relationships suffer. If we go long enough without sleep, we may hallucinate or otherwise act irrationally.

So, sleep we must. But most of us either don't get enough sleep or don't get the right kind of sleep. Often, we are not even aware of it.

This chapter will help you understand your individual sleep cycle and the inner rhythms that make up that nightly cycle. Just as knowledge of your body's waking rhythms can help you improve your life, understanding the similar changes that happen while you sleep can make a big difference in how you feel.

Your Sleep Cycle: A Quiz

• Do you fall asleep easily during the day while reading, watching TV, or doing other sedentary activities?

• Do you find that you are irritable and short-tempered for no particular reason during the day?

• Do you need an alarm clock to awaken you in the morning?

• Do you wake up feeling sluggish and sleepy?

• Do you need a nap to keep you alert through the afternoon and evening?

• Do you regularly "sleep in" an hour or more on weekends?

If you answered yes to any of these questions, you may not be getting enough sleep to meet your individual sleep cycle needs.

• Do you stay awake in bed long after the lights are out, waiting for sleep to come?

• Do you awaken in the morning before your alarm clock goes off?

• Do you spend the last hour or two in bed alternating between sleep and wakefulness?

If you answered yes to any of these questions, you may be trying to get more sleep than your individual sleep cycle demands.

• Do you smoke cigarettes or drink alcohol or a caffeinated beverage late in the afternoon or evening?

• Do you fall asleep with the radio, television, or lights on?

• Do you take sleeping pills?

• Do you sleep in a very cold or very hot room?

• Can outside noises (such as airplanes or street traffic) be heard in your bedroom at night?

• Are you depressed, anxious, or worried?

> *If you answered yes to any of these questions, you may be damaging your natural sleep cycle and not be getting the quality of sleep you need to feel and perform at your best.*

HOW MUCH SLEEP DO YOU NEED?

When Mike got hired by a prestigious East Coast law firm, he knew he'd be working long hours. Yet he promised his wife, Deborah, that he would still find enough time for her and their new son, Joshua. For a while, Mike was able to keep his promise. In the evenings, he would wait until his wife and son were in bed before he took out his briefcase. Frequently, he stayed up until 2:00 A.M. or later to finish his work. On weekends, he would make it a point to rise early and spend as much time as possible with his family before going into his study or driving down to the office to work.

Most weeks, Mike was averaging only about five and a half hours of sleep each night. "I don't need any more than that," he told his wife. But, in fact, he did need more. Lack of sleep soon began to take its toll on him, both physically and mentally. He became more irritable, frequently bickering with his wife about unimportant things, and he was less patient with his son's crying or demands for attention. Mike's work also began to suffer; he made more errors and found himself fighting fatigue at important meetings.

Finally, out of pure exhaustion, Mike decided to cut back on his work load and get more sleep. Within days, his mood and productivity improved.

Some people look upon the daily desire to sleep as a human frailty, an unnecessary luxury that lures too many people away from productive work. Thomas Edison was one of these people. He believed that people could will themselves to sleep less—and that those who could not were simply weak in character. "Most people overeat one hundred percent and oversleep one hundred percent because they like it," he wrote. "That extra hundred percent makes them unhealthy and inefficient. The person who sleeps eight or ten hours a night is never fully asleep and never fully awake."

Legend has it that Edison got by on two hours of sleep each night. His own writings, however, indicate that he slept four or five hours every twenty-four hours. But Edison was also a habitual napper, and thus his reckoning of the amount of time he spent sleeping may not have been accurate.

Do we need to sleep as much as we do? Or, as Edison suggested, can we will ourselves to sleep less? The answer is: It depends. Each of us has an internal sleep/wake rhythm that, like it or not, dictates how much sleep we need each night. It appears to be a rhythm that we are born with and that is interconnected with our other internal clocks. Unfortunately, this is a rhythm we cannot change.

So, how much sleep do you need? First, forget the books that cite the need for eight hours of sleep each night. The amount of sleep needed is different for each individual. But, *on average,* a newborn baby sleeps about seventeen to eighteen hours each day; a four-year-old, about ten to twelve hours; and a ten-year-old, about nine to ten hours. During adolescence, the amount of time we spend sleeping continues to decrease until, by the time we are young adults, we are sleeping an average of seven and a half hours each night. During the next few decades, our sleeping time continues to decline, but gradually—to about six and a half hours a night by old age.

But these are just averages. Each of us has a precise sleep requirement that is genetically determined. We can shave about one hour off that time over a long period with apparently few effects on our physical and mental health, but that does not change the requirement. It just means we can go through the day slightly sleepy and not be aware of it. If we cut back on our sleep by more than an hour, however, we will feel it. We become irritable. Our mental skills and physical abilities deteriorate. We have trouble staying awake during the day—particularly during our usual slump time in the early afternoon.

Today, *most of us get too little sleep*—and we have Thomas Edison, the man who insisted we all sleep too much anyway, to thank for it! For, with Edison's invention, the electric light, we can now extend the day into the night. From movies to meetings, from baseball-under-the-lights to late night television, a host of activities tempt us from our beds—and our sleep. Even children are affected. A survey conducted in the 1960s showed that schoolchildren slept one hour less than children in the early 1900s.

As a result, many of us are walking around out of sorts and out of sync with our internal clocks—and with each other.

GOING WITHOUT SLEEP

Despite the legendary tale of Perseus, whom the Romans supposedly killed by depriving him of sleep, the likelihood of actually dying from a lack of sleep is extremely slim. Only one such case has been reported in modern medicine. It involved a fifty-three-year-old Italian man who, because of a degenerative brain disease, gradually lost the ability to sleep. His thoughts eventually became disoriented, and he lapsed into a stupor, unable to perform even simple tasks. His speech became unintelligible and his movements erratic. His physical health also deteriorated, causing him to develop a lung infection that would not respond to medication. Nine months after he lost the ability to sleep, he died.

This case is very rare. Most of us, no matter how hard we try, cannot stay awake indefinitely. Our bodies will force us to sleep, if only in brief "microsleeps" lasting a few seconds.

Nor is it likely that going without sleep will drive you insane—although it could cause temporary hallucinations. Perhaps the most famous case of someone "losing his mind" through loss of sleep involved New York disk jockey Peter Tripp, who went without sleep for two hundred hours while broadcasting from a booth in New York's Times Square, in January 1959. Toward the end of his sleepless vigil, Tripp began to imagine that a doctor's tweed coat was made of furry worms and that flames were shooting out of a desk drawer. During the night hours, he also experienced periods of paranoia, when he claimed people were trying to slip drugs into his food to force him to sleep.

After the sleepless marathon was over, Tripp slept for thirteen hours. Upon awakening, Tripp's grip on reality completely returned, although he complained of feeling slightly depressed for several months afterward.

Six years later, a seventeen-year-old California high school student, Randy Gardner, stayed awake for 264 hours and 12 minutes (about eleven days) as a science project. Although he became slightly more irritable after the fourth day, Randy experienced none of the hallucinations that had haunted Tripp. Some of his mental skills also stayed remarkably intact. Although he could not recite the alphabet toward the end of his experiment without making an error, he could beat a doctor in one hundred straight games of baseball pinball!

Randy fell asleep within two minutes of completing his experiment (but not before he had answered questions intelligently at a press conference). He slept for fourteen hours and forty minutes, awoke feeling refreshed, and reported no depression or other negative aftereffects.

These and other experiments have indicated that prolonged sleep loss affects each of us differently. Older people, alcoholics, people under stress, and people who are not in good physical shape tend to react more severely to sleep loss.

RHYTHMS OF THE NIGHT

Imagine yourself stepping into an elevator on the top floor of a four-story building. You slowly ride it down, stopping from five to fifty minutes at each of the three lower floors. Then, after about ninety minutes, you make a quick trip from the bottom floor back up to the top floor. But when the elevator doors open, you discover it is not the top floor at all, but a mysterious new floor that you missed on the way down. It is a floor where everything looks strangely familiar, yet unfamiliar at the same time. After exploring this new floor for about ten to fifty minutes, you get back in the elevator and begin your descent once more.

Each night, as you sleep, you take a similar elevator ride as you travel up and down through the four different stages of sleep. A complete round-trip ride usually lasts from sixty to ninety minutes, and is repeated four or five times each night. It is one of our clearest ultradian rhythms.

Stage 1 (light sleep): 5 to 10 percent of total sleep time

• Muscles relax, accompanied by a floating or drifting sensation.

• Brain waves slow down from an average waking speed of thirteen to thirty-five pulsations per second (known as *beta waves*) to eight to thirteen pulsations per second (*alpha waves*).

• Blood pressure falls.

• Pulse rate slows by about ten beats per minute.

• Blood sugar and calcium levels rise.

• Body begins detoxification process, excreting toxins from cells. This process usually peaks (for people who sleep at night) around 4:00 A.M., which is also when body temperature falls to its lowest point of the day.

• We can be easily awakened during this stage—indeed, some scientists don't consider this stage true sleep. If awakened, we may deny being asleep.

Stage 2 (light sleep): 50 percent of total sleep time

• Brain waves slow even further to four to eight pulsations per second (*theta waves*), with periodic rapid bursts of activity known as spindles.

• Body metabolism—blood pressure, temperature, pulse—continues to decline.

• Eyes roll slowly from side to side; if eyelids are lifted, we do not see.

• We can be easily awakened during this stage of sleep. If awakened, we may deny being asleep.

Stage 3–4* (deep sleep): 25 percent of total sleep time in young adults (more for children, less for older people)

• Long, slow brain waves of less than four pulsations per second (*delta waves*).

• Breathing becomes even and muscles fully relax.

• We are not easily awakened during this stage; only a loud noise or repetition of our name will pull us out of our slumber.

• If awakened, we feel groggy and confused—children especially.

• This is stage when bedwetting, sleepwalking, and sleeptalking are most likely to occur.

• Body's metabolism reaches its lowest point; however, a few hormones are released during this stage of sleep, including those that help the body grow and heal.

Stage 5 (REM, or dream, sleep): 20 to 50 percent of total sleep time

• Characterized by rapid eye movements (REM).

• Dreaming occurs.

• Brain waves quicken to waking speed of thirteen to thirty-five pulsations per second.

• Heartbeat and blood pressure become irregular, sometimes fluctuating wildly.

• Need for oxygen increases, so breathing becomes faster.

• Adrenal glands begin pouring larger amounts of hormones into the body.

• Steroids increase to highest levels of the day.

• Body experiences a sleep paralysis; if awakened, we may not be able to move for a few seconds.

• Increase of flow of blood to genitals causes penile erections in males of all ages.

---

*In the past, scientists used to divide deep sleep into two separate stages. This differentiation is no longer made, but the numbering continues to reflect past practice.

Our first nightly encounter with REM sleep typically lasts from five to fifteen minutes. We then sink back into Stage 3–4 sleep, although this time our deep sleep does not last as long as before. Then, we climb upward again into another REM sleep period, where we spend a longer time dreaming than the first time around.

And so it goes through the night: four or five complete cycles, each lasting approximately sixty to ninety minutes. Each time the deep, or Stage 3–4, sleep gets shorter—until, by the end of the night, we may skip it altogether. Each time the REM sleep grows longer, until, by the last cycle, we may be spending fifty minutes or more dreaming.

This explains why we tend to remember dreams more just before rising in the morning. At that time of our sleep cycle, we are simply spending more time dreaming—and thus are more likely to awaken during a dream.

The Sleep Instinct

Every living creature, from the elephant to the mouse, from the Beluga whale to the butterfly, spends part of each twenty-four-hour cycle resting. Butterflies fold their wings at night; birds tuck their heads under their wings. Lizards, lobsters, and turtles become immobile. Even fish rest for a time each day, their breathing slowed, their coloration less intense. Many species of fish spend their resting period on the bottom of their watery home; others prefer a spot closer to the surface. Sea otters also snooze in the ocean, wrapping themselves in kelp to keep from drifting ashore.

Resting, however, is not always the same as sleep. Only mammals and birds experience true sleep, with its alternating periods of dreaming and deep sleep. Scientists speculate that warm-blooded animals evolved into sleepers as a way of reducing body temperature and conserving energy when food supplies were scarce.

THE CRITICAL STAGES:
DEEP SLEEP AND DREAM SLEEP

Of all the stages of sleep, deep sleep (Stage 3–4) and dream sleep (REM) are most essential to health and well-being. Indeed, people who function well on three hours or less of sleep each night—spend most of their sleeping time in these two stages and little time, if any, in Stages 1 and 2, indicating that the need for light sleep is not strong.

What Deep Sleep Does for You

• *Restores body and brain.* Many scientists believe deep sleep acts as a kind of nightly tune-up for the body, restoring the brain and body by helping to heal wounds and rebuild tissue lost during the day. Just how sleep accomplishes this is not known, however. Deep sleep may also help fight bacterial and other infections by allowing the body to devote all its resources to fighting the infection—which may explain why we become sleepy when we are ill.

• *Stimulates growth.* The body's growth hormone is released during deep sleep, usually early in the nightly sleep cycle. The release of this hormone is essential to growth, which may explain why the total amount of time we spend each night in deep sleep decreases as we get older and stop growing. Toddlers, for example, spend 20 to 30 percent of their sleeping time in Stage 3–4; but college students spend only 10 to 15 percent in Stage 3–4. By the time we are sixty, we devote only 1 to 2 percent of our total sleep to deep sleep. This is one reason older people don't need as much sleep as younger people.

The release of growth hormone during sleep may also help explain why children who are abused often fail to grow at normal rates. Fear and forced awakenings frequently keep these children from getting enough deep sleep early in their nightly sleep cycle. When taken out of their homes and placed in a safe place where they can sleep undisturbed, normal growth begins again, often at a spectacular pace.

Parents of growing children should make sure their children have a quiet, secure place to sleep. Once your child is in bed, turn down the volume on your television set or stereo, so the child's sleep will be undisturbed. This is especially important during the first few hours of your child's sleep, when most of the growth hormone gets released. Make the time right before going to bed as pleasant and unrushed as possible, so your child will have a deep and unworried sleep.

• *Maintains mental health.* We also may need deep sleep for our mental health. During one study, subjects were deprived of deep sleep for seven consecutive nights. By day seven, the subjects could perform tasks at the same level of competence as before the study, but they had experienced profound psychological changes. They became withdrawn, less aggressive, and complained of vague physical ailments. These symptoms disappeared when the deep sleep was restored.

It may not be surprising, therefore, that researchers have discovered that people who are depressed spend less time in deep sleep and more time in light sleep (Stages 1 and 2) than other people their age. They also tend to enter their first REM sleep quickly—within thirty to fifty minutes of falling asleep. And they spend more time in REM sleep during the first third of the night than during the last third—a complete reversal of the normal pattern. (To find out more about the connection between sleep rhythms and depression, see chapter 4.)

What Dream Sleep Does for You

• *Consolidates and sorts memory.* Some scientists believe dreaming helps consolidate memory by filing away all the day's experiences into the appropriate pigeonholes in the brain. Others believe it acts as the brain's housecleaner, sorting the day's experiences into those to be kept (remembered) and those to be thrown away (forgotten).

• *Enables learning.* We may also need REM sleep to help with brain development and learning. This theory stems partly from the

Things That Can Disrupt Your Sleep Rhythms

The following factors can play havoc with your natural sleep patterns by making your sleep more shallow or by awakening you altogether. In either case, the disruption robs you of necessary deep sleep and REM sleep:

• Dieting (Loss of weight can cause more frequent awakenings during the night.)
• Caffeine
• Alcohol
• Low-sodium diet (elderly people only)
• Cigarettes
• Sleeping pills
• Exercising vigorously—if you're not accustomed to it
• A sleeping partner who snores or thrashes about in the night
• A dog or cat moving about in your bedroom
• Sporadic, unpredictable noises, such as airplanes flying overhead or street traffic
• A room that is too warm (more than 75° F) or too cool (less than 60° F)
• Hunger
• Stress

fact that REM sleep makes up almost all of a baby's sleep in utero and half of a baby's sleep for about six months after birth. This is the time when the brain is doing its most explosive growth.

In addition, several studies have shown that people who are involved in learning something new or adapting to a new situation during the day increase their REM sleep that night. One study, for example, observed the sleep patterns of people who had suffered brain damage and who were being coached to improve their speech. Those whose speech improved increased their REM

Things That Can Enhance
Your Sleep Rhythms

The following factors can help you make your sleep patterns more regular and thus ensure that you get enough deep and REM sleep:

• Regular aerobic exercising (at least twenty continuous minutes)—particularly in the late afternoon
• Sex right before bedtime
• Following a bedtime ritual
• Relaxation exercises right before bed
• A bedtime snack high in carbohydrates, such as toast and jam
• A dark and quiet bedroom
• A "white noise," such as the hum of a fan or the sounds on tape of ocean waves, to mask other, sleep-disturbing noises

sleep, while those whose speech failed to improve had no such increase.

Because most of our REM sleep comes during the last few hours of sleep, cutting sleep short in the morning may actually interfere with learning. Students who get up early in the morning to deliver newspapers or train for a sports activity should be encouraged, therefore, to go to bed early. An earlier bedtime will help ensure that they get a complete night's sleep—and enough REM sleep to help them absorb and retain what they learned in school the day before.

• *Ensures our physical safety.* Yet another theory suggests that, because we usually wake up briefly after dreaming, REM sleep is nature's way of ensuring our physical safety. According to this theory, if we spent the entire night in deep sleep, we might be oblivious to dangers—such as predators—around us.

Whatever the reason, we need to dream. Experiments have shown that when we cut short our REM sleep by awakening, we quickly return to it when we fall asleep again—strong evidence of our physical need for REM sleep. In fact, sleep researchers who try to keep people out of the REM stage have to awaken their subjects every few minutes.

It is probably not true, however, that we need to dream to keep from going crazy, as sleep researchers used to believe. Early studies showed that depriving people of their dreams led to anxiety, irritability, and difficulty in concentrating. If the dream deprivation was prolonged, some people even hallucinated. But, because the researchers were having to awaken their subjects almost constantly, it is believed these findings were the result of losing *all* sleep, not just dream sleep.

MEETING YOUR INDIVIDUAL SLEEP NEEDS

To feel your best, it is important that you have regular sleep habits. This means going to bed at the same time each night and waking up at the same time each morning. It also means sleeping for the full length of time required by your biological clock.

If you keep your sleep habits regular, you'll be surprised how refreshed and alert you'll begin to feel, even after only a few days of the new schedule. To get on the schedule that's right for you, however, you'll need to determine two sleep factors: how long you should sleep and your ideal bedtime.

How to Determine How Long You Should Sleep

Free-running—sleeping and waking when you feel like it—is the most reliable way to find out how many hours your body needs to devote to sleep during each twenty-four-hour cycle. For most of us, free-running is possible only during vacations, when we don't have the time-constraining demands of a nine-to-five job. For parents, this also means a vacation away from young children, who can cut sleep short in the morning with cries for breakfast and other requests.

You'll need a minimum of five days to determine your personal requirement for sleep. The longer you allow yourself to free-run, the more accurate your findings. Each day, you should go to sleep when you feel tired and get up when you spontaneously awaken. Do not oversleep or nap. During the free-running experiment, use a sleep diary to keep track of when you fall asleep, when you awaken, and the total number of hours you sleep each day. (You'll find an example of a sleep diary on page 65.) At the end of the experiment, add up all the totals—except for days 1 and 2—and divide by the number of days. This will show your ideal sleep length time. (Totals from the first two days are not included, because most of us will sleep extra-long those days to make up for the sleep we have deprived ourselves of while on our regular schedules.)

Remember, most people have a twenty-five-hour circadian cycle, so you may find both your bedtime and your wake-up time drifting later each day while you're free-running. But once you have stopped free-running and are back to the time constraints of your everyday duties, it is important that you keep to a strict bedtime and wake-up time to ensure that you continue to get your optimum amount of sleep.

How to Determine Your Ideal Bedtime
First, determine your ideal wake-up time. At what time do you need to get up in the morning to get your day started properly? 6:00 A.M.? 7:00 A.M.? 9:00 A.M.? Now count backward from that time the number of hours of sleep you have determined you need nightly. That will give you your ideal bedtime.

To keep to your ideal bedtime, however, you must adhere to a consistent wake-up time. Oversleeping—even for an hour—will push back your sleep/wake cycle and make it more difficult to fall asleep on time that night. Soon you'll be off your schedule, losing sleep and feeling less than your best when awake.

If you must catch up on your sleep, do it by going to bed earlier, not by sleeping later. It will cause less confusion for your biological clocks.

HOW TO STAY UP LATE AT WEEKENDS AND STILL AVOID THE MONDAY MORNING BLUES

Paul, a twenty-three-year-old computer programmer, enjoyed the night life. A year earlier, during his last year of college, Paul and a group of his friends would study almost every evening until 10:00 P.M., at which time they would shut their books and head for a local nightclub, where they would drink and dance until closing. Paul made sure that he had no early morning classes, so he could catch up on his sleep in the morning.

When Paul got a nine-to-five job, however, he soon realized he could no longer keep up his late night lifestyle and get to work on time and fully conscious. He decided to limit his carousing to Friday and Saturday nights. After all, he reasoned to himself, he could make up the lost sleep on Saturday and Sunday mornings.

Paul's plan worked—except for one small hitch. No matter how hard he tried, he found it difficult to go to bed at a reasonable hour on Sunday night. As a result, he always felt tired on Monday and sometimes on Tuesday, as well.

The Monday morning blues—that feeling of tiredness we experience on Mondays after a particularly active weekend—are directly related to our sleep/wake cycle. Staying up late, sleeping in, and snoozing on a Sunday afternoon in a backyard hammock or on the beach can turn our circadian cycles inside out. By Monday morning, we are feeling out of sync and short of sleep. The result: the blues. Often, this feeling of tiredness continues into Tuesday and Wednesday, as our bodies try to reset to the weekday schedule.

The ideal way to avoid the Monday morning blues is to stick to your regular sleep habits. But many of us, like Paul, enjoy staying up at weekends, getting together with friends and family members for a party, a play, or a late night pizza.

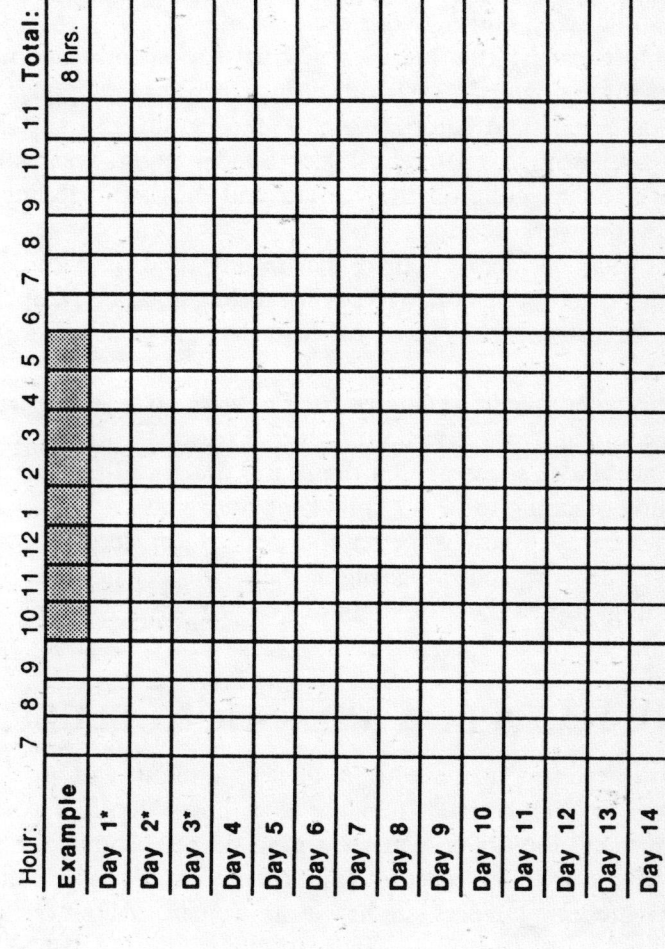

Hour:	7	8	9	10	11	12	1	2	3	4	5	6	7	8	9	10	11	Total:
Example			▒	▒	▒	▒	▒	▒	▒	▒	▒							8 hrs.
Day 1*																		
Day 2*																		
Day 3*																		
Day 4																		
Day 5																		
Day 6																		
Day 7																		
Day 8																		
Day 9																		
Day 10																		
Day 11																		
Day 12																		
Day 13																		
Day 14																		

*Do not include these days in any of the calculations

Fortunately, there is another way to avoid the Monday morning blues. Stay up as late as you like, but *make sure you get up at your usual wake-up time*. In other words, *don't* sleep in. And don't take naps during the day. Then, when night rolls around again, you'll find yourself able to fall asleep at or before your regular bedtime, and you'll avoid pushing your cycle ahead.

Unfortunately, this means you'll have to spend your waking hours during Saturday or Sunday (or perhaps both, if you stay out late on Friday and Saturday nights) feeling sluggish and sleepy. But, by getting a full night's sleep Sunday night, you will keep your circadian cycle relatively intact and thus avoid feeling tired during the work week.

You may decide, of course, that feeling tired on Saturday and Sunday is too high a price to pay for feeling good on Monday and choose, instead, to sleep in on Saturday and Sunday mornings. That's a decision only you can make.

This technique of forcing yourself to wake up at your usual time is also useful if you lose sleep for other reasons—perhaps because you are under stress or you drank too much caffeine late in the evening or caring for a sick child kept you awake part of the night. Remember: It is always better to make up lost sleep by going to bed earlier the following night than by sleeping later in the morning. It puts the least strain on your circadian rhythms.

ARE YOU A LONG OR SHORT SLEEPER?

If you need less than six hours of sleep a night, you are what is known in sleep research parlance as a short sleeper; if you need more than nine hours, you are a long sleeper. About two out of ten people are short sleepers (slightly more men); about one out of ten are long sleepers (slightly more women). About one out of twenty-five people sleep less than five hours or more than ten.

Perhaps the shortest sleeper verified in a sleep laboratory was a seventy-year-old English nurse who averaged sixty-seven minutes of sleep a night. She said she had slept that way since she was a child and could never understand why other people wasted so much time sleeping!

Contrary to common opinion (including Thomas Edison's views on the matter), sleeping less is not a prerequisite for success. One survey of 509 "men of distinction" revealed an average sleep-length time of 7.4 hours. And Albert Einstein, a man who was certainly no slouch, reportedly slept twelve hours a day.

Long sleepers and short sleepers may have different personalities, however. According to one study, short sleepers tend to be nonworriers who are generally satisfied with themselves and their lifestyles. Long sleepers, on the other hand, tend to worry more, both about themselves and the world. They also tend to be nonconformists, artistic, and to suffer more from anxiety and mild depression. Both long and short sleepers get the same amount of deep sleep; but long sleepers also get more REM sleep. Scientists speculate that they may need that extra REM sleep to restore the brain and psyche.

How a Bad Night's Sleep Affects Your Athletic Performance

Tossing and turning at night can churn up your biological rhythms and detrimentally affect how you run, cycle, or swing a racquet the following day. Generally, the more demanding an athletic endeavor, the more it will be affected by a loss of sleep. Thus, a marathon runner will have his or her performance suffer more as a result of a bad night's sleep than, say, a golfer. This is especially true if the sports event is scheduled at a time of day when performance rhythms are already low, such as during the early morning hours.

If you have an important athletic event coming up, be sure you get an adequate amount of sleep. Having your rhythms in sync will help your body perform at its best.

These findings are not conclusive. Indeed, another study has shown that there is no difference in personality, intellectual ability, or physical health between long sleepers and short sleepers.

WHEN RHYTHMS GO AWRY

As we saw in chapter 1, most of us tend to have a daily circadian rhythm that is slightly longer than the twenty-four-hour cycle imposed on us by the sun. Still, we find it relatively easy to make the daily adjustment to the shorter cycle. We continue to go to bed at a reasonable hour and to wake up in time to join the rest of society at the office or factory by 9:00 A.M.

For some people, however, this daily adjustment to a twenty-four-hour solar rhythm is difficult. They have a natural sleep/wake cycle that runs too long or too short, or one that changes from day to day. For these people, following society's schedule is a difficult, if not impossible, task. They often want to sleep when others are awake, and vice versa.

If you are one of these people who suffer from an out-of-kilter circadian rhythm, take heart. Sleep researchers have developed ways to manipulate sleep habits to overcome these problems—although the process takes time and discipline.

If You Have Trouble Falling Asleep

All of us have trouble falling asleep from time to time. Daily stresses, too much caffeine, or even an unfamiliar bed can keep us counting sheep into the wee hours of the morning.

But for natural night owls—people with body temperatures that rise and fall later than those of other people—the inability to fall asleep at a reasonable hour is a common and frustrating occurrence. Night after night, they spend hours tossing and turning; often it's three or four in the morning before they finally fall asleep.

These people are not insomniacs. They have no problem sleeping soundly for the necessary hours, once they fall asleep. It's the *timing* of their sleep cycle that is off, not its quality. The peak in their body temperature simply occurs at a later-than-average time in the day. As a result, their periods of alertness and sleepiness are not in sync with those of the rest of society.

Sleep scientists have a name for this problem: delayed sleep phase syndrome. Fortunately, there is a cure—a process called *chronotherapy,* or time therapy. It works on the theory that it is much easier to reschedule sleep by staying up later than by going to bed earlier. Patients are required to delay their bedtime by two or three hours each night until they have reset their bedtime at a desirable hour. Then, they must strictly keep to that bedtime—or risk throwing their rhythms off again. This can have an adverse effect on their social ife, but most patients report that it's worth cutting back on their night life so they can enjoy the company of people—and hold down a job—during the day.

For example, if you regularly fall asleep at 3:00 A.M., on the first night of your chronotherapy treatment, you would stay awake until 6:00 A.M. Then you would sleep for your usual amount of time, using an alarm to make sure you don't oversleep. On the second day of the program, you would go to bed at 9:00 A.M.; the third day, at noon; the fourth day at 3:00 P.M.; until, by the seventh day, you reach midnight. That would become your regular bedtime.

Chronotherapy has worked for many people. It is best undertaken, however, under the care of a sleep specialist, for if the process is interrupted for any reason before it is completed, your sleep pattern could become more chaotic than before.

If You Have Trouble Staying Awake

A much less common problem, but still a frustrating one for those who suffer from it, is advanced sleep syndrome. People with this problem have body temperatures that rise and fall at earlier-than-average points of the day. As a result, they fall asleep and wake up earlier than they would like—sleeping, for example, from 9:00 P.M. to 4:00 A.M. These are the larks or morning people who rise before the sun, but who set with it also. They nod off at parties, fall asleep in movie theaters, and are generally regarded as dead wires after 8:00 P.M.

Because they can get up in time to go to work, people suffering from advanced sleep syndrome find that their sleep pattern doesn't usually interfere with their job. But it can interfere with their social life, and, for this reason, many people with the syndrome would like to readjust their sleep habits to become more in step with the rest of society.

If you suffer from this syndrome, chronotherapy can help. The therapy, however, will not be as drastic as that involved with delayed sleep phase syndrome. Simply force yourself to stay awake fifteen minutes longer, then keep that later bedtime for a week. The next week, push your bedtime ahead another fifteen minutes; the third week, yet another fifteen minutes; and so on until you reach your desired bedtime. Once you reach your goal, be sure to stick with it, or you may find yourself slipping back into old habits.

If You Have Trouble Keeping a Regular Bedtime

Some people are night owls, some people are larks, and still others are both. People who are alternately larks and owls have weak circadian rhythms. Their body temperatures show no regular pattern, rising and falling at varying times during successive days. Some days they may be on a twenty-five-hour circadian cycle, some days on a twenty-seven-hour cycle, and other days on a twenty-three-hour cycle. As a result, they cannot plan when they will sleep and when they will be awake, for the pattern shifts from day to day.

People with this problem often use sleeping pills, pep pills, and naps to help them function during the day and sleep at night. Unfortunately, this only makes the problem worse.

If you suspect that you suffer from a weak or irregular circadian rhythm, you should contact a sleep disorder clinic to have the condition verified. Chronotherapy can then be used to help you establish a regular sleep cycle. The exact pattern the chronotherapy will take will vary depending on your individual case.

When Chronotherapy Fails

Some people's abnormal circadian rhythms are so strong (or so weak) that they cannot adjust to the twenty-four-hour day, even after chronotherapy. For these people, finding a lifestyle that fits their particular rhythm is a better solution than trying to adopt a new rhythm. One woman journalist, for example, became a freelance writer after attempts at keeping a regular schedule proved impossible. Many night owls have decided to take night jobs when attempts to become "day people" have failed.

However, before you give up on rescheduling your sleep/wake cycle, be sure you've given it a good try. Besides keeping strict sleep habits, you should make sure your other activities, such as exercising and eating, are regular, too.

Sleep-Disturbing Drugs

Caffeine. Caffeine can play havoc with your sleep/wake cycle. And with good reason. Caffeine is chemically related to amphetamines and is a powerful stimulant to the central nervous system.

Caffeine can be found in a variety of products—coffee, tea, chocolate, soft drinks, and many over-the-counter pain-relieving medicines. If you want to avoid it, read labels carefully. Caffeine's "buzz" usually peaks two to four hours after it is taken. Drinking a cup of coffee or having a caffeine-containing soft drink in the evening, therefore, may not only make it more difficult for you to go to sleep, but it may also cause you to wake up frequently during your sleep. Caffeine can also make you excessively sleepy during the day once the "buzz" wears off.

Alcohol. Once recommended in moderate doses as a sleep-promoter, alcohol in any amount is now recognized as a sleep-disturber. For, although a small nightcap can relieve tension and make it easier to fall asleep, once the alcohol becomes metabolized the body begins to experience a mini-withdrawal, which can cause frequent awakenings during the night and a feeling of irritability and sluggishness in the morning. Alcohol also suppresses REM sleep. Because your body will try to make up for the lost REM sleep, your sleep the night *after* a night of drinking is therefore likely to contain more dreams. This is why alcoholics often experience nightmares and disturbed sleep for weeks or months after becoming sober. Sometimes the brain damage caused by alcoholism is so severe that a return to normal sleep patterns is impossible.

Nicotine. Nicotine is a powerful stimulant that can interfere with sleep. Heavy smokers have a more difficult time falling asleep and spend less time in both deep and REM sleep than nonsmokers, according to studies conducted in sleep labs. They also wake up more during the night. In one study, eight smokers abruptly gave up cigarettes after puffing on more than one and a half packs a day for at least two years. After going cold turkey, the exsmokers fell asleep faster and woke less during the night. So, add a good night's sleep to all the other health benefits of not smoking.

Sleeping pills. Some people take sleeping pills for years, even decades, believing the pills improve their sleep. In reality, sleeping pills lose their effectiveness after several weeks, and the sleep they do give during that period is of poor quality. Sleeping pills "lighten" sleep, resulting in less REM sleep and less deep sleep. They can also cause you to wake up feeling tired and lethargic. If you want to quit your sleeping pill habit, do so slowly and under the care of a physician. Sudden withdrawal after long use can create severe sleeplessness, as well as nightmares, as your body tries to make up for lost REM sleep.

CAUGHT NAPPING: GOOD OR BAD?

For some people, naps can be a wonderful pick-me-up, even better than the traditional coffee break. Many eminent people including Napoleon Bonaparte, President John Kennedy, and Winston Churchill were all famous nappers. In fact, Churchill was almost rapturous on the subject. "Nature had not intended man to work from eight in the morning until midnight without the refreshment of blessed oblivion, which, even if it lasts only twenty minutes, is sufficient to renew all vital forces," he wrote.

Churchill may have been right, for recent research seems to indicate that naps are part of nature's plan for us. Many scientists now believe that we were meant to sleep twice a day, once during the afternoon and again at night. These are the times of day when outside temperatures are at their extremes and, therefore, when the body needs more energy to carry out its activities. By sleeping at these times, we enable our bodies to conserve precious energy. As previously noted, in many hot climates, everyone stops at midday to take a daily siesta and retreat from the heat.

In countries where daily siestas are not part of the culture, most regular nappers are very young, very old, or living in institutions. Surveys have also shown, however, that a surprisingly high number of college students—about 50 percent—nap regularly. Some people nap as a psychological boost, to relieve stress or tension. Others nap simply to replace lost sleep.

Napping may even be beneficial before a big test or activity requiring physical skill. In one study, college students who regularly napped showed improved scores on skill and memory tests taken just after a nap. Students who were not used to napping, however, awoke feeling sluggish and scored poorly on the tests—so don't try a nap before a big exam unless you're a regular napper!

Napping is not without its risks. It may cure daytime drowsiness, but it may also interfere with your regular nightly sleep rhythm, making it difficult to fall asleep or making the sleep shallower and less efficient. That's why sleep researchers usually recommend that poor sleepers avoid naps.

The time of day you take your nap, the length of the nap, and whether or not you are a regular napper will all influence the quality of a nap—that is, whether or not it refreshes you. You will wake up less groggy and more refreshed from a morning or early afternoon nap (if your regular sleep time is during the night) than from a nap taken in the late afternoon. This is because morning naps are made up mostly of REM sleep and late afternoon naps of deep sleep, and it is easier to wake up from REM sleep. However, falling asleep is another matter. You'll find it more difficult to drift into a morning nap than into an afternoon one, in large part because the brain's alertness chemicals are at their peak in the morning. Also, those ninety-minute periods of "sleepability" that we experience throughout the day are shorter during the morning than in the afternoon.

Scientists have also discovered that short naps (twenty minutes or less) are just as revitalizing as long ones—and less disruptive to regular sleep rhythms. In addition, regular nappers wake up more refreshed than those who take only an occasional nap. When naps are part of the regular daily rhythm, they are less disruptive.

If you feel you must take daytime naps to keep your energy levels high throughout the day, then try to schedule your naps during early or midafternoon, when your alertness levels experience a natural slump and you are most vulnerable to sleepiness. By napping at that time of day, you will also wake up less groggy than you would if you napped in the late afternoon. Select a quiet, comfortable, and darkened room in which to sleep, and use an alarm clock to limit the nap to twenty minutes or less. A high-carbohydrate snack eaten right before your nap time can help induce sleep by triggering the production of calming chemicals in the brain.

To break a napping habit, try to do something active—such as taking a walk—during the time of day you would normally be napping. Avoid places or situations where you could easily close your eyes and fall asleep, such as concerts, movies, or watching television. Until the habit is broken, you may wish to use a caffeinated beverage, such as coffee, tea, or cola, to help keep you awake. A high-protein meal or snack can also help keep you awake by releasing alertness chemicals from your brain. (For more about how to use food to keep you awake or help you sleep, see chapter 6.)

THINGS THAT GO BUMP IN THE NIGHT

Many of the things that happen to us while we sleep are tied to a specific stage of sleep. We will discuss some of the more common problems and experiences of sleep, and when they are most likely to occur.

Stages 1 or 2	Deep Sleep	REM Sleep
Tooth grinding	Night terrors	Nightmares
Head banging	Sleepwalking	Sleep paralysis
Myoclonic jerks	Bedwetting	Painful erections
	Sleeptalking	Cluster headaches
		Sleep-related asthma
		Hypnagogic hallucinations

Tooth Grinding (Bruxism)

Researchers estimate that between 5 and 20 percent of adults grind their teeth in their sleep. No one is sure exactly what causes tooth grinding, but it may be triggered by stress or by a physical or neurological disorder. The problem should be treated, for it can severely damage teeth and gums. Cures for tooth grinding range from teeth-clenching exercises to hypnotism to wearing a rubber or plastic mouth guard over the teeth during sleep.

Head Banging

More common among children than adults, rhythmic head banging or body rocking usually occurs at the onset of sleep. For some children, head banging may be a symptom of emotional stress; but with most children, it is nothing more than a comforting habit (like thumb sucking) and disappears as the child grows older.

Myoclonic Jerk

Sometimes, during the first stage of sleep, a tiny burst of neural activity occurs in the brain, causing the body to jerk awake suddenly. This experience is common and normal.

Night Terrors

Unlike nightmares, which occur during a REM stage of sleep, night terrors usually occur during the first deep sleep (Stage 3–4) of the night. During a night terror, the sleeper often screams and sits up in bed, terrified. Breathing and heartbeat are rapid, eyes appear glazed, and the sleeper is difficult to console. Even if severely shaken, the sleeper may take several minutes to awaken, and will then usually be unable to remember what image caused the terror. Night terrors are more common in children than in adults—perhaps because children spend more time in Stage 3–4 early in the night. In children, night terrors do not reflect a psychological problem, although stress can be a factor. The episodes usually disappear by adolescence. In adults, night terrors can be a reflection of severe, chronic anxiety. Keeping regular sleep habits can help reduce the need for long periods of Stage 3–4 sleep early in the night and thus reduce the incidence of night terrors.

Sleepwalking (Somnambulism)

Lasting from a few minutes to a half hour or more, sleepwalking occurs when a sleeper awakens only partially from deep sleep (Stage 3–4). Most sleepwalkers remain in bed; they may simply sit up abruptly and look about them with a glazed, unfocused stare. Some, however, get out of bed and wander about their homes or neighborhoods—sometimes dressing, eating, or even driving a car. Their movements are usually clumsy and can threaten their safety. Sleepwalkers have been known to mistake a window for a door.

Sleep researchers estimate that 15 percent of all children sleepwalk at least once, and 1 to 6 percent of adults do it regularly. In children, sleepwalking is considered normal and does not reflect emotional problems, although stress, as well as fatigue and a high fever, can sometimes trigger an episode. Most children stop sleepwalking by the age of fifteen. In adults, sleepwalking is usually caused by extreme stress or an emotional trauma.

Keeping regular sleep habits can reduce the incidence of sleepwalking by decreasing the amount of time spent in Stage 3–4 sleep early in the night. Psychotherapy may also be helpful for adults who sleepwalk.

Bedwetting (Enuresis)

Although bedwetting can occur during any stage of sleep, it happens most frequently during deep sleep (Stage 3–4) and during the first third of the night. Bedwetting is common in children, especially in boys, and tends to run in families. Although many theories exist, the cause of bedwetting in children remains a mystery. Most experts agree, however, that it is usually not related to emotional or physical problems. Bedwetting that begins in adults, however, may be caused by urinary tract disease, diabetes, epilepsy, or sleep apnea (a serious condition in which the sleeper stops breathing for episodes of ten seconds or more) and should be brought to the attention of a physician.

Sleeptalking (Somniloquy)

Lady Macbeth's soliloquy to the contrary, sleeptalking generally involves only the briefest burst of words, and usually nonsensical ones at that. Frequently, sleeptalking takes the form of a simple sound—a grunt, laugh, shriek, or groan. Children talk in their sleep more than adults, but, in either case, it is not considered a problem that needs treatment.

Nightmares

Known as dream anxiety attacks by sleep scientists, nightmares are terrifying dreams that cause us to awaken short of breath, sweating, and in great fear. Unlike the night terrors discussed earlier, nightmares occur during the REM stage of sleep and usually during the second half of the night, when REM periods are longer and more intense. Stress often triggers a nightmare, as does withdrawal from alcohol or other drugs.

Sleep Paralysis
During REM sleep, the brain's locus coeruleus cells, which are located near the brain stem and which help the body's muscles keep their taut tone, become inactive. As a result, our muscles become limp and we experience a kind of sleep paralysis. Scientists believe this paralysis occurs to prevent us from physically reacting to our dreams and possibly hurting ourselves. Cats who have had their locus coeruleus cells removed, for example, will jump up and chase imaginary animals or objects while dreaming.

Usually we remain blissfully unaware of this paralysis; but sometimes we may awaken abruptly from REM sleep to the frightening fact that our muscles will not move for a few seconds, or even minutes. If this should happen to you, don't panic. Just relax and wait for your muscle tone to return. Blinking or moving your eyes often helps break the paralysis. If the paralysis or a feeling of weakness in the muscles lasts longer than a few minutes, however, consult a doctor. Persistent sleep paralysis can be a symptom of other medical conditions, such as thyroid disease, potassium deficiency, and narcolepsy (a sleep disorder characterized by uncontrollable and sudden sleepiness).

Movement During REM Sleep
In some people—mostly older men—the locus coeruleus cells do not become inactive during REM sleep. (See Sleep Paralysis, above.) As a result, their muscles do not become limp while they dream; on the contrary, they move about while they dream, punching, kicking, sometimes even rising from their beds and walking or running around.

This tendency to be active during REM sleep is known as REM behavior disorder. It differs from sleepwalking, which occurs during deep sleep. The cause of REM behavior disorder is not known, although studies have shown that one-third of the people who suffer from it have an underlying neurological disease. People who are physically active while dreaming should consult their physician for a neurological checkup.

Painful Erections

It is natural and normal for males of all ages to have erections during the REM stages of their sleep. These erections are not caused by sexy dreams, as is commonly believed, nor are they caused by a full bladder. They occur because one of the natural rhythms of the REM stage of sleep is an increased flow of blood to the genitals, which engorges and enlarges the penis. The average young male adult has four erections each night, totaling about 191 minutes, or one-third of his total sleep time. The average man in his seventies, on the other hand, averages three erections a night, for a total of 96 minutes, or one-fifth of his sleep time.

For some men, these sleep erections can be painful. The cause is often a physical one, such as a temporary blockage in the blood vessels leading to or from the penis or, in uncircumcised men, a foreskin that is too tight. Men with painful erections should consult a urologist for treatment.

Cluster Headaches

The increased flow of blood through the brain during REM sleep may be the cause of the painful cluster headaches that plague some sleepers. The term cluster comes from the tendency of this kind of headache to come and go in periodic bursts. The headaches usually occur repeatedly over a period of several weeks or months and then stop for an equal or longer period. Cluster headaches first begin after adolescence, and more men than women suffer from them.

The typical cluster headache starts about one to two hours after the onset of sleep, when REM periods are in full swing. The headache usually attacks just one side of the head, often resulting in one teary eye, one stuffed nostril, and a reddening of one side of the face. The sleeper may awaken at the onset of the headache or may feel it only upon awakening in the morning. Getting up, walking around the room, or even doing vigorous exercises sometimes relieves the pain. REM-suppressing drugs can also help in some cases. Contact your physician for treatment.

Sleep-Related Asthma

Asthma is the swelling and narrowing of the body's air passages from the windpipe to the lungs. It results in wheezing, coughing, and shortness of breath. REM sleep seems to exacerbate asthma attacks, but scientists are not sure why. It may be because the nervous system becomes excited during dreaming. Or it may be because it's easier for the body to respond to an asthma attack during REM sleep. Supporting this last theory are studies that show asthma patients get less deep sleep than other people. The attacks may disturb their deep sleep without actually awakening them.

Hypnagogic Hallucination

Sometimes people fall directly into REM sleep, without going through the other stages of sleep first. When this occurs, the beginning images of a dream may seem real, for the dreamer may not be fully asleep yet. These dreamlike images are known as hypnagogic hallucinations. They are often frightening—of an animal or person about to attack, for example—and, if accompanied by sleep paralysis, can be terrifying. Some people believe that these nondream images and sounds are the source of religious visions and out-of-body experiences.

Hypnagogic hallucinations are most common among people who suffer from *narcolepsy,* a sleep disorder characterized by excessive daytime sleepiness. People with narcolepsy fall asleep easily and usually drift immediately into REM sleep. People who do not suffer from narcolepsy can also experience these hallucinations, particularly if they have gone without sleep for twenty-four hours or more, or are withdrawing from amphetamines or other drugs that suppress REM sleep. Hypnagogic hallucinations are also more likely to occur during a nap, particularly one taken in the morning, when REM sleep predominates.

Tips for Improving Your Sleep Cycle

As we've seen in this chapter, paying attention to your sleep cycle can improve your physical and mental health. Here is a summary of tips for ensuring that your nighttime rhythms add to your daytime health and happiness.

• Assess your sleep needs and determine the optimum number of hours you need to sleep. (See sleep diary on page 65.)

• **Keep regular sleep hours—even at weekends.**

• If you stay up late, be sure to get up at your regular time the next morning.

• Avoid alcohol, cigarettes, and caffeine—especially after 6:00 P.M.

• Do not use sleeping pills.

• Use naps judiciously. If you nap, do so regularly. Never nap if you have trouble sleeping at night.

• Avoid falling asleep with the light or radio on. Although the noise or light may not awaken you completely, it can rob you of needed deep sleep.

• Do at least twenty minutes of aerobic exercise daily— preferably in the late afternoon.

• Make sure you don't shortchange yourself on sleep. Remember, most people—and that probably includes you—are not getting enough sleep!

Four

~

The Measure of Your Moods

Sorrow breaks seasons and reposing hours,
Makes the night morning, and the noontide night.
—WILLIAM SHAKESPEARE
Richard III

Every autumn, starting around mid-October, Jane, a young mother and part-time librarian, notices a change in her personal habits. She begins to eat more, and often feels sleepier than usual. She even begins to take an afternoon nap with her two-year-old daughter on her days off from work—something she never does in the summertime. She also notices that her mood is darker, especially when she awakens in the morning, and that it takes a considerable effort to drag herself out of bed in the morning. All of these symptoms persist until early April, when warmer weather and longer days seem to lighten her mood and her cravings for food and sleep.

Joseph is a forty-eight-year-old engineer for a large midwestern computer company. Since his late teens, Joseph has known that his temperament is not the best early in the morning. In fact, he can be downright rude if asked a question before 9:00 A.M. As the day progresses, however, Joseph becomes charming and accommodating. Fortunately, Joseph's company has a flex-time policy that allows employees to set their own work hours as long as they put in a full day. Joseph has chosen the ten-to-six shift. Now he need not worry about offending his coworkers early in the day.

All of us are aware of our emotional ups and downs—the times when we feel happy and confident and the times when we're down in the dumps. What most of us don't realize, however, is that—like Jane and Joseph—our mood swings often occur at regular, and even predictable, intervals.

These cycles are most obvious—and thus most studied—in people with serious mental illnesses, such as manic-depressive psychosis. But minor cyclic shifts in mood—daily, monthly, and seasonal—in healthy people also have recently been recognized and studied.

This chapter will help you understand the biological basis for your changing moods. It will also help you learn how to recognize your individual cycles and how you can help yourself feel better when you are on the down swing of a cycle.

Your Mood Vulnerability: A Quiz

• Do you do a lot of traveling across time zones within relatively short amounts of time?

• Does your job require that you rotate shifts regularly?

• Do you have trouble keeping a regular sleep schedule?

• **Do you work nights during the week, yet at weekends** shift to a more normal "day" life?

• Do you tend to stay out and sleep in much later than usual on the weekends?

• Are you caring for a newborn baby in your home?

• Do you rarely get outside during daylight (sunlight) hours?

If you answered yes to any of these questions, you may be more vulnerable to rhythm-related depression.

THE RHYTHM OF BLUES

As we saw in chapter 2, our sense of well-being is generally strongest during the late morning hours. These daily changes in our emotions are subtle—so subtle that we usually are unaware of

	1	2	3	4	5

Depressed,   Happy,
sad   elated

	Day 1	Day 2	Day 3	Day 4	Day 5	Total	Average
6 a.m.							
7							
8							
9							
1 0							
1 1							
1 2							
1							
2							
3							
4							
5							
6 p.m.							
7							
8							
9							
1 0							
1 1							
1 2							

them. For example: Around 11:00 A.M., a work assignment that seemed impossible to do when it was handed to you the previous afternoon may suddenly begin to look entirely doable. Or, the behavior of your child—behavior that made you want to tear out your hair in frustration the evening before—may now seem manageable, maybe even a bit endearing.

More obvious are the monthly swings in mood associated with the menstrual cycle of women. In general, menstruating women feel best during the first half of their cycle and worse during the second half, just prior to menstruation. Preliminary evidence also seems to point to similar monthly emotional cycles in men. (See chapter 5 for a discussion of these monthly cycles in both men and women.)

Our moods also appear to ride a seasonal roller coaster, with winter being the "down" season and summer the "up" one. As long winter nights come upon us, we tend to turn inward and become grumpier and more melancholy.

These mood cycles appear to be connected to the biological rhythms within us. For most of us, these cycles are a normal and acceptable part of our lives, and we are able to adjust to them without too much trouble.

Sometimes, however, the biological rhythms affecting mood can go askew. A particular rhythm might get out of control, for example, rising *too* high or falling *too* low. Or a rhythm or group of rhythms may break away from each other and from the synchronizing influence of the sun. The result: moodiness, anxiety, and, in severe cases, depression.

Tracking Your Daily Moods

Because our emotions are so strongly affected by outside events, it is often difficult to find a consistent pattern to our daily moods. However, if you rate how you are feeling every hour for several days, you should begin to see an underlying rhythm to your emotions. Use the five-point scale shown opposite and record each reading in the spaces provided.

For best results, you should continue the test for a minimum of three days. Then, add up the hourly totals and divide by three (the number of days of the test) to determine your average hourly reading. Plot that average reading on the chart opposite.

What to look for. Generally, our moods peak around four hours after we awaken, which means they are better in the morning than in the evening—even for people who are evening types. People who are suffering from clinical depression, however, have mood rhythms that are just the opposite: They feel worse in the morning, best at night.

This does not mean, of course, that everyone who has a reversed mood cycle is clinically depressed. But, if your moods do seem to be topsy-turvy, as shown on the chart, *and* if you are experiencing some of the other symptoms of depression, such as loss of sleep, extreme fatigue, and feelings of worthlessness, you may wish to consult a physician or psychologist.

STRESS AND THE BLUES: A BREAK IN THE BEAT

Each day, we have to deal with some kind of stress. It may be something relatively mundane, such as missing a bus, or it may be life-shattering, such as the death of a loved one.

A stressful event is not always a negative one. Positive stress, such as a job promotion or the birth of a baby, can also bring tension, strain, and even depression into our lives.

The relationship between stress and depression is extremely complex and appears to vary according to the person and the situation involved. But researchers are beginning to see a distinct connection between stress and our biological rhythms—a connection that may go a long way toward understanding how stress affects our moods.

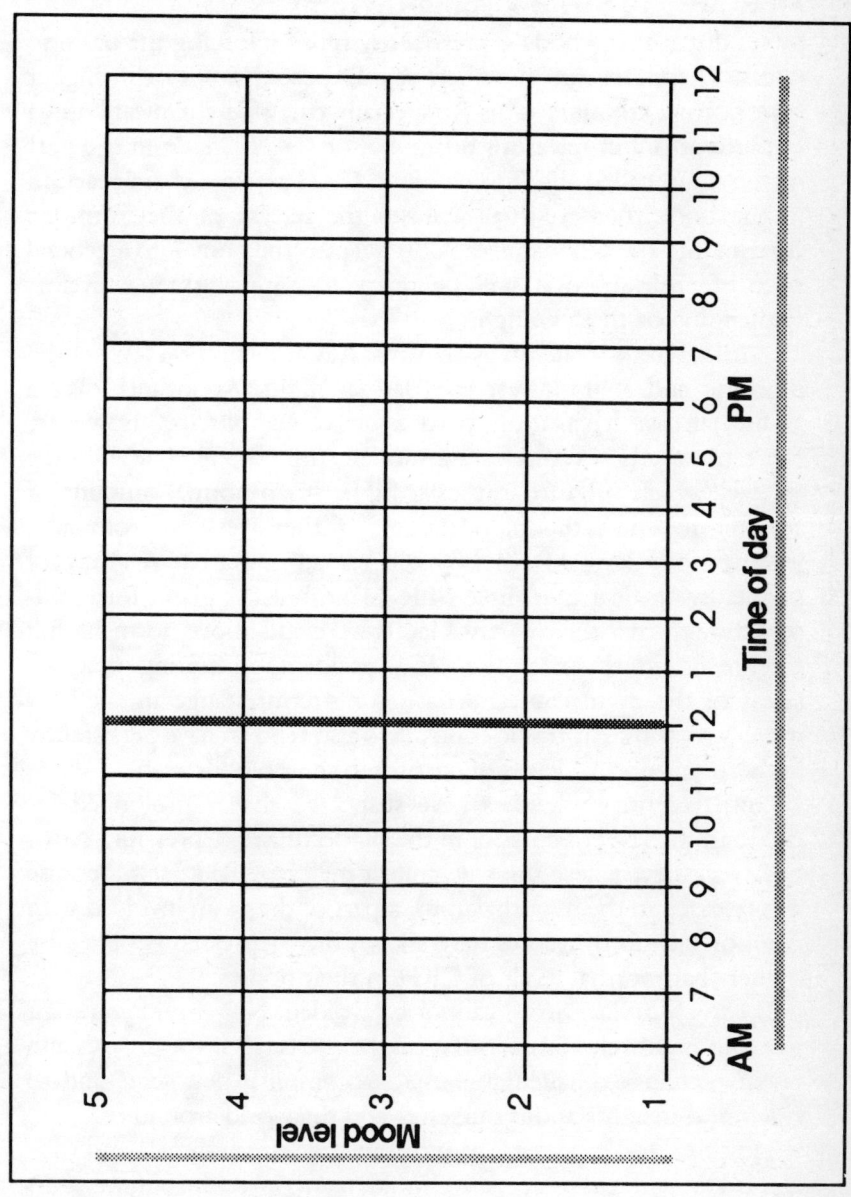

How Stress Disrupts Your Rhythms

Stress disrupts the body's internal rhythms by forcing the brain to release chemicals into the body at unexpected times or in higher than normal amounts. This is especially true of a chemical known as corticotrophin-releasing hormone, or CRH. Made in the part of the brain called the *hypothalamus,* CRH starts a chain reaction in the body that eventually causes the release of the hormone cortisol into the bloodstream. Cortisol puts the body into a general state of arousal, so it will be ready to cope with whatever is challenging or threatening it.

Cortisol has a steady daily rhythm: It is at its highest level in the morning and at its lowest level late at night. Annoyance over a traffic jam or frustration with a broken appliance, however, can cause the brain to release a sudden surge of CRH at any time of day, which, in turn causes a higher-than-normal amount of cortisol to enter the bloodstream. If the stress is prolonged, however, the level of cortisol will remain elevated. A study of people living near the Three Mile Island nuclear plant, for example, showed that their cortisol levels were still above normal a full seventeen months after the 1979 accident, indicating that the stress of the event caused a basic long-term change in the hormone's rhythm. Air traffic controllers also tend to have persistently higher-than-normal levels of cortisol in their bloodstream.

Until recently, scientists have suspected that a prolonged increase in the level of cortisol in the blood causes depression. Now, however, they believe the true culprit may be CRH itself, because it is released in the hypothalamus, a part of the brain involved with emotions. Indeed, studies have shown that depressed people have higher-than-normal levels of CRH in their bodies.

What you can do. Stress cannot be avoided. Nor would you want to avoid it entirely. Stress often can be the catalyst for growth and positive change, challenging us to accomplish new feats and to reach new insights about ourselves and the world around us.

However, to keep stress from throwing your rhythms—and your emotions—off balance, you should learn how to minimize its physical effect on your body. You can do this in a variety of ways. For example:

• *Practice relaxation techniques.* These include deep breathing exercises, yoga, biofeedback, transcendental meditation, progressive muscle relaxation, and autogenic training (the use of repetitious phrases and imagery to relax). Many popular books have been written on these and other stress-reduction techniques. Look these books over, then choose the method that appeals to you. Your family doctor should be able to direct you to classes or individual professionals in your community who will be able to help you get started with the technique or techniques of your choice.

• *Pace yourself.* Whenever possible, avoid scheduling too many stressful events close together. Spread work deadlines across many days or weeks, for example. Also, avoid attempting too many major life changes at the same time—such as moving into a new home right after the birth of a baby.

• *Follow your rhythms.* As you plan your day-to-day routine, keep in mind those times of the day when you are best at certain activities (see chapter 2), and schedule your activities accordingly. You'll find it less stressful, for example, to work on your taxes at noon, when your mind is best equipped to handle complicated thought processes, than at 9:00 P.M., when it's not. Women should keep track of monthly changes in their ability to cope with stress due to their menstrual cycles.

• *Listen to your body.* Don't ignore the telltale signs of stress— sleeping difficulties, indigestion, and fatigue, for example. When you feel as if you are pushing your body too hard, slow down, relax, and take some time to enjoy the small, pleasurable things in life.

• *Select your fights.* You don't want to keep anger bottled up inside you, but you also don't want to let every little hassle turn into a stressful confrontation. Learn when to fight and when to walk away. If, for example, you get less than satisfactory service at a restaurant, rather than demanding to see the restaurant's manager and creating a scene, just tell yourself that it doesn't really matter and that next time you will simply take your business elsewhere. Also, learn how to give in once in a while, whether it be in a discussion about politics or in a dispute with your spouse about which movie you're going to see that night. You don't always have to be right.

• *Be with people you like.* Spending time with people you care about can be a great stress-reducer. It also helps to talk with trusted friends and relatives during times of stress in your life. Work to develop your personal support group and turn to it when the need arises.

• *Exercise.* A regular exercise program can go a long way toward easing the physical effects of stress. Contrary to what many people believe, exercise enhances rather than drains your energy. When done at the same time each day, it can be a strong zeitgeber, helping to reset other rhythms. The exercise program must be regular, however; sporadic physical activity, such as an occasional game of tennis or football, or an activity that is not appropriate to your physical condition, can be a harmful stressor.

How a Disruption of Rhythms Causes Stress

Just as stress can disrupt your rhythms, so can a disruption of your rhythms cause stress. Sometimes our rhythms get thrown off for obvious reasons—jet lag, shift work, or the middle-of-the-night feeding demands of a newborn infant. At other times, the source of the disruption is more subtle.

This was the case with Phyllis, a twenty-eight-year-old woman who recently had separated amicably from her husband of five years. Soon after the separation, Phyllis began to notice small changes in her physical well-being. To begin with, she was having trouble sleeping. She was used to falling asleep while her husband watched late night talk shows or sports events on the television in their bedroom. Now the room seemed too quiet, and she had difficulty falling asleep.

Her digestive system also seemed out of whack. Her husband

had done most of the cooking in their home, and they had always made a point of eating breakfast and dinner together. Now, with her husband gone, Phyllis found herself eating irregularly, and the food she put on her breakfast and dinner plates was more often processed than home-prepared.

Besides a lack of sleep and indigestion, Phyllis began to notice a more general feeling of malaise. She often felt fatigued and depressed for seemingly no reason at all. She couldn't focus very long on projects at work. And she seemed to catch every cold and flu virus around. Perhaps, Phyllis told herself, she was more upset about the separation than she thought.

While emotional stress was no doubt contributing to Phyllis's complaints, there was something else going on within her. Phyllis was experiencing a desynchronization of her biological rhythms caused by the loss of an important zeitgeber—her husband. When her husband left, so did many of the important social time cues that helped Phyllis's internal rhythms stay in sync with themselves and with the outside world. Her body's digestive system, for example, was thrown off cue when it no longer received the same amounts of food at the same time each day. Her sleep/wake cycle experienced a similar upheaval with the absence of the noise cue (television) for falling asleep.

Eventually, of course, Phyllis's rhythms will find new social time cues. In the meantime, the disruption of the rhythms is causing stress to both her body and mind.

Few of us think of the people we live with as time cues, but that's exactly what they are. They help us set our daily routine, which is extremely important in keeping our rhythms in sync and our bodies and minds healthy.

The kind of internal desynchronization experienced by Phyllis is even more severe in a person who loses someone through death. Not only is the death emotionally devastating, leading to acute depression, but the loss also means an abrupt change in the daily routine of the survivor. Without well-established time cues, the survivor's biological rhythms are thrown off and must begin to reset themselves to new cues. That reshuffling of rhythms can, in turn, cause the normal depression of bereavement to become even deeper.

It should be noted that not everyone who experiences a desynchronization of rhythms becomes notably depressed. The degree to which desynchronization affects our moods seems to vary from person to person. Chronobiologists believe that a genetic predisposition may be at work, making some of us more prone to rhythm-related mood swings than others. People who have extroverted personalities, for example, appear to be better able to handle desynchronization than people with more introverted temperaments.

What you can do. Try to keep the beat of your internal rhythms steady by sticking as close to a regular routine as possible. This means eating, sleeping, and exercising at about the same time each day. If you must change your routine, do it as gradually as possible, so that your rhythms have time to adjust.

During times of extreme stress, it is even more important to keep to as regular a schedule as possible. This, of course, is often difficult to do. But even doing something as simple as making sure your meals are scheduled at their regular times can go a long way toward easing rhythm-related stress.

WINTER BLUES

Just as we get stress-related blues from time to time, we suffer winter blues as well. As the days shorten and the nights lengthen, we tend to oversleep, overeat, and, in general, take on a slightly darker view of the world. These feelings linger through the winter, gradually lifting in the spring, as the days lengthen.

For most of us, this annual change in mood is mild and does not interfere with our lives. For some, however, the effects can be dramatic and devastating. People who normally sleep five or six hours during the summer may find themselves sleeping twice as long in the winter. In extreme cases, they may be so overwhelmed with feelings of gloom and lethargy that they find it impossible to leave their homes from December until February.

These people are suffering from seasonal affective disorder (SAD), a type of depression that only recently has been recognized and studied by scientists. The exact number of people suffering from this illness is not known, but judging from the response researchers receive when they advertise for subjects with SAD symptoms, it seems to be widespread.

SAD sufferers typically begin to sink into their depression around December, and the illness persists until March or April, with January and February being the worse months. Spring usually brings a complete recovery, including an almost euphoric surge of energy.

Women seem to be more susceptible to SAD than men, and the illness also occurs most often in families that have a history of other kinds of depression. The symptoms can begin as early as childhood, although most people do not recognize the cycle until they are in their twenties or thirties. The symptoms are also more severe in dark, foggy climates than in sunny, bright ones. One physician who suffers from SAD discovered, for example, that his symptoms increased when he moved from Texas to Baltimore, and they became even worse when he moved a few years later to Philadelphia.

The key to understanding SAD seems to be a hormone called melatonin, which is secreted into the bloodstream by the pineal gland, a little-understood pine cone–shaped organ inside the brain. Melatonin's daily rhythm is similar to that of legendary vampires: It appears at the onset of darkness and begins to disappear at the break of dawn. In other words, darkness triggers the chemical's release; sunlight suppresses it. It should come as no surprise, therefore, that melatonin reaches its seasonal peak in the body during the dark winter months.

Scientists are not sure what role melatonin plays in the human body, but it appears to be involved intimately in the regulation of the body's internal rhythms. In animals, it seems to regulate seasonal behavior, such activities as hibernation, migration, and breeding. Without melatonin, animals are unable to keep track of the seasons and will hibernate or reproduce at unsuitable times of the year.

Chronobiologists theorize that melatonin may also cause a kind of seasonal hibernation in humans—a hibernation that produces winter depression.

Fortunately, there is help for SAD sufferers. Exposure to bright, *full-spectrum* lights for an hour or two each morning or evening

---

Let There Be Light

Sunlight plays an important role in our daily and seasonal moods. Here are a few tips to help you make sure you get enough light:

• Expose yourself to direct sunlight for at least fifteen minutes each day.

• Use your sunglasses sparingly, especially in winter. For it's through your eyes, not through your skin, that sunlight suppresses melatonin.

• If you work inside all day, try to position your desk or work space as close to a window as possible.

• If you spend a great deal of time inside, either at home or at work, consider using special full-spectrum fluorescent lights that mimic natural sunlight instead of other artificial lights. Although SAD sufferers need an intense treatment of full-spectrum light to lift their symptoms, others have found that lower doses of these special lights can help lift milder cases of the winter blues.

Full spectrum lights are available under the brand name True-Lite. Write to Full Spectrum Lighting Ltd, Unit 4/5, Wye Industrial Estate, London Road, High Wycombe, Bucks, HP11 1LH (Phone: 0494 448727).

appears to lift the spirits of people with this kind of depression. These special lights mimic the light from the sun. By artificially creating a summer's day in the middle of winter, they apparently are able to fool the body into producing less melatonin, thus relieving the depression.

To maximize your energy during winter or gloomy weather, you should increase your exposure to the powerful zeitgeber of light. (See box on page 94.) However, if you suspect you have SAD, you should not attempt self-treatment. Discuss your symptoms with your doctor and raise the possibility of receiving light therapy. If your doctor is not familiar with this type of therapy, you can refer him or her to the SAD Support Group, 51 Bracewell Road, London W10 6AF.

How SAD Differs from Other Serious Depression

Many of the symptoms of SAD—low self-esteem, a loss of interest in everyday activities, increased lethargy—are similar to the symptoms of other types of serious depression. But SAD's symptoms are also significantly different. Here's how:

SAD	Other Depression
Depression occurs during winter months	Depression typically occurs in spring or autumn
Depressed mood lifts in spring	Depressed mood typically lifts in summer
Appetite increases during winter, particularly for carbohydrates	Appetite usually decreases
Time spent sleeping increases	Time spent sleeping decreases

THE CYCLES OF DEPRESSION

SAD is not the only kind of serious depression that advances and retreats in cycles. Clinical depression—a severe mood disorder that is typified by intense feelings of sadness and often accompanied by serious disturbances in normal bodily functions—is also a recurrent illness. (For a more detailed description of clinical depression, see the box on page 97.)

Sometimes the cycles of clinical depression are short and, thus, easy to spot. This is especially true of manic-depressive psychosis, a type of mental illness in which depressive episodes alternate with periods of mania, or excessive excitement, hyperactivity, and rapidly changing thoughts. For example, several cases of people with forty-eight-hour manic-depressive cycles (episodes of depression and mania falling on alternate days) have been reported in the medical literature. In one famous case, a salesman in Washington, D.C., became so morose and apathetic on his depressed days that he was unable to leave his car once he arrived at the offices of his clients. On his "up" days, on the other hand, he was the ideal salesman—loquacious and aggressive. He worked around his illness by making appointments with clients on alternate days.

Such short cycles are relatively rare, however. Clinical depression is more often characterized by annual cycles. It is most common in spring. Not surprisingly, spring is also the peak season for suicides.

Researchers have found that the circadian rhythms of depressed people are different from those of nondepressed people. For example:

• Body temperature tends to peak earlier in the day in depressed people than in nondepressed people.

• The daily rise of cortisol begins earlier than usual in depressed people—specifically, early in the evening rather than late at night.

• Levels of the thyroid stimulating hormone (TSH), which indirectly affects mood, fall to their lowest point at night rather than in the morning, as is the case in nondepressed people.

• The daily peak of the hormone melatonin occurs earlier in the night in depressed people than in nondepressed people.

All this seems to point to the depressed person's body clocks being out of kilter. And, indeed, that is just what scientists have

---

When the Normal Blues Become Depression

All of us feel sad from time to time. Usually it is only a transient state, and we are able to pass through it without disrupting our normal daily routines. For some people, however, the feelings of sadness, remorse, or guilt become acute and severely interfere with normal life. These people are suffering from clinical depression, an illness that afflicts millions of people in today's society.

The boundary between normal feelings of sadness or grief and the abnormal condition of depression is not always clear. However, psychiatrists and others who treat people with depression have come up with the following common symptoms for clinical depression. Generally, four of these symptoms must persist for at least two weeks to be considered a clinical depression:

• Eating too much or too little
• Sleeping too much or too little
• Too much restlessness or too much lethargy
• Loss of interest or pleasure in usual activities, including decrease in sex drive
• Overwhelming fatigue
• Feelings of worthlessness and self-reproach
• Feelings of diminished ability to think or concentrate
• Recurrent thoughts of death or suicide, or actual suicide attempts

Depression and Christmas

For decades, the popular press—and many therapists—have been proclaiming Christmas as the most depressing time of the year. It was believed that the stress of Christmas activities, such as shopping for presents, plus the usually unfilled hope of having a perfect holiday, surrounded by loving family and friends, inevitably led to dark feelings of depression, loneliness, and hopelessness.

Recently, however, this depressing view of the Christmas season was revealed to be nothing more than a holiday myth. A seven-year study at a major university medical center revealed that psychiatric emergencies are less frequent during the Christmas season than at other times of the year.

Another mistaken belief is that suicides also peak around the Christmas holidays. However, statistics collected by the US National Center for Health Statistics show that more suicides occur in April than in any other month. In fact, the fewest occur in December. And more people are admitted to mental hospitals during the summer months than at any other time of the year.

found. Using sophisticated new wristwatch devices that can monitor a person's rest and activity levels over a long period of time, researchers have discovered that the circadian rhythms of many seriously depressed people free-run, much like the rhythms of volunteers who spend long periods of time isolated in underground caves.

Specifically, the body's two basic pacemaker systems—the one governing the body temperature cycle and the one governing the sleep/wake cycle—become out of sync with each other. The temperature rhythm, the stronger of the two, usually continues on a roughly twenty-four-hour cycle, while the sleep/wake rhythm goes off entirely on its own.

Scientists believe it may be this uncoupling of the body's rhythms that triggers many of the symptoms of clinical depression. How, or why, is not yet known.

THE SLEEP CYCLE AND DEPRESSION

The most common complaint among seriously depressed people is that they have trouble sleeping. Dick, a forty-five-year-old university professor, is a good example. When in one of his depressive episodes, he finds it extremely difficult to fall asleep, and when he finally does drift off, his sleep tends to be light and not very refreshing. He frequently wakes up around 4:00 A.M., long before his alarm clock is set to go off, feeling tired and anxious. He often feels an overpowering sense of gloom at that hour.

Doctors used to believe that this kind of sleep difficulty was simply a result of the anxiety and dark thoughts that come with depression. As we've seen, it is now thought that the sleep problems that depressed people experience appear to be caused not by dark thoughts, but by a free-running sleep/wake cycle, one that is no longer in sync with other internal rhythms or with the twenty-four-hour cycle of the sun.

In other words, the depressed person's normal sleeping hours have been rearranged. Sleepiness no longer occurs at the time of day (usually night) when the body temperature cycle is at its lowest point. Instead, sleepiness occurs at longer and longer intervals—thirty, forty, or even fifty hours apart.

It's not surprising, therefore, that depressed people take longer to fall asleep and wake up earlier than nondepressed people. Only part of their rhythms are geared to sleeping every twenty-four hours.

A long, free-running sleep cycle is not the only sleep characteristic that changes when serious depression hits. The *kind* of sleep seriously depressed people experience is also much different from that of nondepressed people. Here are some of the basic character-istics that differ:

• *Fragmented sleep.* Depressed people travel through the various stages of sleep more quickly than nondepressed people. For exam-ple, instead of passing through the complete sleep cycle four times a night, a depressed person may make the trip eight times.

• *Inefficient sleep.* Depressed people have less deep (Stage 3–4) sleep and more shallow (Stages 1 and 2) sleep than nondepressed people. They also tend to awaken more easily during the night. As a result, they may spend more time in bed than nondepressed people, but less time actually sleeping.

• *Earlier REM sleep.* REM, or dream, sleep occurs earlier in the night for depressed people. Instead of entering REM sleep about ninety minutes after falling asleep, a depressed person will begin to dream within thirty to fifty minutes. (Insomniacs who are *not* depressed also enter REM sleep earlier than normal, but usually no sooner than sixty minutes after first dozing off.) Depressed people also experience more dreaming during the first third of their sleeping period than during the last third—exactly the oppo-site of nondepressed people.

This isn't to say that everyone who has trouble sleeping is depressed. You can have insomnia and not be clinically depressed. In one study of people who were referred to sleep disorder centers, only about one out of six were diagnosed as having depression.

Nor do all depressed people suffer from insomnia. People suffer-ing from SAD, for example, usually spend an excess amount of time sleeping. In all, about one out of five people diagnosed with a serious depression tend to oversleep rather than undersleep. People who oversleep when they are depressed also tend to overeat. The younger you are, the more likely you will respond to stress by oversleeping and overeating. Conversely, the older you are, the more likely you will cut back on your sleep and your eating during a bout of depression.

MANIPULATING SLEEP TO TREAT DEPRESSION

For most of us, being deprived of sleep has a negative effect on our moods. If we get too little sleep, our temperament usually takes a turn for the worse the next day. We may snap at our spouse over breakfast or complain about a client during a morning conference at work or bemoan the fate of the world at dinner. Sleep deprivation makes us feel downright grumpy, lethargic, or irritable—maybe even melancholy. So, for most of us, getting a good night's sleep can often be just what the doctor ordered for a blue mood.

But, if we are suffering from serious depression, a *bad* night's sleep may actually be the medicine we need! Instead of making a clinically depressed person more melancholy, sleep deprivation has been found to lift the sufferer's dark feelings of apathy, hopelessness, and sadness—at least temporarily. The reason: When non-depressed people lose sleep, they throw their circadian rhythms out of sync, which can lead to a more depressed state of mind. But, as we've seen, the rhythms of depressed people already are off kilter. For them, staying up all night may get their temperature and sleep rhythms back in sync. The synchronization usually occurs somewhere between 1:00 A.M. and 5:00 A.M. After this, their mood notably improves.

Unfortunately, the good effects of skipping a night's sleep go away as soon as the depressed person falls asleep again. However, scientists have found a way to achieve longer-lasting results. First, they keep a depressed patient awake for an entire night. The following evening, they have the patient go to bed very early in the evening—around 6:00 P.M. Six hours later, the patient is awakened and told to stay awake until the following evening.

This therapy shifts the patient's sleep/wake and temperature cycles back into synchronization with one another. Unfortunately, the effect does not last; the cycles eventually uncouple. But it can produce remissions of several weeks—long enough for antidepressant medications to begin working in the body. *A warning:* Do not try this therapy without the help of a trained professional; when done improperly, it can intensify rather than relieve the depression.

Many psychiatrists are now using a twofold treatment of sleep deprivation and antidepressant drugs to treat clinical depression. Interestingly, the drugs and the sleep deprivation appear to do the same thing—shorten the abnormally long sleep/wake cycles of depression.

Tips for Easing the Blues

If you are suffering from a depression that is prolonged or seriously interfering with your life, you should seek professional counseling and care. However, if your depressed state is due to a mild case of the blues, you can take several rhythm-related steps to brighten your mood:

• Keep as regular a daily routine as possible.

• Make sure you get a full, restful night's sleep. This means keeping a regular bedtime and sleeping for as long as you need (see chapter 3).

• Get outside in the sunshine as often as possible—at least fifteen minutes each day.

• Use special high-spectrum lights for your indoor lighting wherever possible.

• Minimize the desynchronizing effect of stress on your body's rhythms by practicing relaxation and other stress-reducing techniques.

Five

Your Sexual Cycles

Never the time and the place
And the loved one all together!
—ROBERT BROWNING
Never the Time and the Place

Soon after they were married, Karen and John decided to use natural family planning as their method of birth control. This meant that Karen had to learn to read the monthly physical changes in her body so she and John could abstain from making love during the few days she was fertile. It was a hassle for Karen and far from foolproof, but she and John chose natural family planning because of their religious beliefs and their concern about the safety of many other forms of birth control.

During the first few months of using the method, Karen was so busy watching for the signs of fertility that she didn't notice another, less obvious, change that was also part of her cycle. Around the fifth month, the pattern became so clear that she couldn't ignore it. It had to do with her sexual desire. She found, to her great annoyance, that she wanted to make love most at the very time of the month when it was forbidden!

Women like Karen who use the natural family planning method of birth control (NFP) often complain that they feel most like having sex at the time of the month when they are at greatest risk of becoming pregnant. In the past, sex researchers might have explained this increased desire as just another example of absence—or abstinence—making the heart grow fonder. Today, however, scientists recognize that an increase in sexual desire during the fertile period of a woman's menstrual cycle is biological rather than psychological. It is yet another example of a biological rhythm.

We have other sexual rhythms as well, rhythms that influence not only how we feel about sex, but also how we feel about all aspects of our lives. This chapter will take a look at these rhythms and show their impact on our work, play, lovemaking, and even on our susceptibility to illness.

Surprisingly, less research has been done in the area of sexual rhythms than in other areas of chronobiology, but scientists are beginning to uncover some fascinating clues to the rhythmic mysteries of sex.

Your Sexual Cycles: A Quiz

- Do you feel sexier in the morning than at night?
- Do you feel sexier in autumn than in spring?
- For women only: Do you feel sexiest at the midpoint of your menstrual cycle?
- For men only: Do you notice that your beard grows more rapidly around the same time each month?

If you answered yes to these questions, you may already be in tune with some of your natural sexual cycles.

I'M IN LOVE.
IT MUST BE—AUTUMN?

If we are to believe poets and songwriters, spring is the time for love—and sex. Nature tells us something different, however. Our sexiest season is autumn, not spring. Testosterone levels in both men and women reach their yearly peak in late summer and autumn. Testosterone is one of the androgens, or so-called male hormones, that have a direct effect on sexual behavior. They are present in women as well as in men, although in smaller quantities.

It's no surprise, then, that most babies are born in summer and early autumn—about nine months after the autumn peak in testosterone. Nature may have intended it this way to ensure that the weather and food supply would be amenable to delivering and nourishing a newborn baby.

Social customs can throw these natural seasonal birth rhythms askew, however. In predominantly Catholic countries, for example, where couples are encouraged to abstain from sex during Lent, births tend to peak in early winter—about nine months after the Easter season.

So, be on the alert in autumn. It appears to be a particularly easy time of the year to get swept into a passionate romance—a romance you may later, under the influence of winter's "cooler" rhythms, regret.

A Time to Be Born

If you are pregnant, plan for a middle-of-the-night trip to the hospital. As any midwife, obstetrician—or chronobiologist—will tell you: More babies are born at night than during the day. Around 20 percent more, to be precise.

Why? It may be that nature wants us to have our babies under the cover of darkness to protect us from animal predators.

The most common birthing hour is between 3:00 A.M. and 4:00 A.M., and the least common one is exactly twelve hours later, between 3:00 P.M. and 4:00 P.M. Late afternoon is the peak time, however, for stillbirths. Scientists are not sure why.

You're also most likely to feel those first labor pains at night. Don't rue your luck, however, if your labor begins around midnight, for nighttime labor tends to be shorter than daytime labor!

YOUR DAILY AND WEEKLY LIBIDO

Besides their seasonal peaks and valleys, androgens fluctuate with some regularity every twenty-four hours. Generally, androgen levels are highest between 8:00 A.M. and 12:00 noon, and lowest between 6:00 P.M. and midnight.

Does that mean that we practice more sex in the morning? Not really. Studies have shown that our most active time for sexual activity is not morning, but the more convenient evening hours—specifically, 10:00 P.M. That may be a good time to fit in lovemaking, but it's a poor one as far as our androgens are concerned.

The time slot second in popularity is 7:00 A.M. But, again, this probably reflects convenience (many couples are still in bed together at that hour) rather than the result of surging hormones.

Social habits seem to play an even stronger role in weekly sexual rhythms. People who work a regular Monday-to-Friday week have sex most frequently on Saturday and Sunday (with Sunday morning being particularly popular). People whose work week differs from this norm tend to increase their sexual behavior on other days—generally the days they are not working. From this, it seems likely that social customs rather than biology determine these weekly rhythms.

THE MENSTRUAL CYCLE

The most obvious—and most studied—of the reproductive rhythms is the menstrual cycle. It is also second only to the sleep cycle (perhaps) in its influence on women—and, indirectly, on the men who live with women.

The menstrual cycle has been enveloped in myth and mystery for centuries. Menstruating women have been blamed for everything from making crops wither in the field to souring wines. Ironically, menstrual blood has been attributed with a variety of beneficial properties: Primitive societies have used it to protect men against battle wounds, to put out fires, and to treat headaches, epilepsy, and other illnesses.

Today, most of the myths about the menstrual cycle have been discarded, but much of the mystery remains. The menstrual cycle is one of the most intricate cycles of the human body and, for many women, one of the most important in terms of their everyday physical and mental health.

For example, where a woman is in her monthly menstrual cycle may affect whether:

- she feels optimistic or pessimistic about life
- she desires sex or could just as well do without it
- she wins or loses in a sports event
- she fights off or succumbs to a viral infection

Despite the profound effect of the menstrual cycle on women's lives, few women understand their cycles enough to recognize how to benefit from the high points of their cycles and minimize any discomfort or distress that occurs during the low points. This understanding is possible, however. All it takes is some general knowledge about the inner rhythms of the menstrual cycle and the compilation of specific data about your individual cycle.

ONCE A MONTH—OR THEREABOUTS

The activity of the monthly menstrual cycle is centered around the tiny sacs of eggs, known as follicles, that are found in the ovaries. A woman has hundreds of thousands of these follicles, but each month only one grows and develops to the point where it releases its egg. Occasionally—although rarely—more than one follicle is released and fertilized, and twins, triplets, quadruplets, or even quintuplets may be born as a result.

Some women can actually feel the egg being released deep within the womb. It is a sensation known as *mittelschmerz*— German for "middle pain," because it occurs at the midpoint of the cycle and is characterized by a temporary aching or cramping in the abdomen.

After the egg breaks from the ovary (ovulation), it is drawn into a nearby fallopian tube and then makes the three-day journey to the uterus, a distance of about four inches. During this trip, the egg may be fertilized if sperm reach the egg and one sperm penetrates it. In fact, the lining of the uterus (endometrium) begins to thicken immediately after ovulation in preparation for receiving and nourishing the fertilized egg. Usually, however, the egg is not fertilized by the time it reaches the uterus. It either disintegrates or is passed out of the body. When a fertilized egg is not received, the built-up lining of the uterus sheds and is expelled—a process known as menstruation.

What Can Delay or Shorten Your Menstrual Cycle?

Your menstrual cycle can lengthen or shorten—or even stop—for a variety of reasons. The most common ones are listed here. Always contact your physician if you notice a pronounced change in your cycle; it may be a sign of a serious health problem.
- Crash dieting
- Prolonged use of oral contraceptives
- Excessive exercise
- Emotional stress
- Climate and time zone changes
- Tranquilizers containing chlorpromazine or meprobamate, such as Equanil, Meprate, or Largactil
- Heavy marijuana use
- Chronic iron deficiency
- A major operation
- Thyroid dysfunction
- Pituitary tumors (small, benign tumors on the pituitary gland in the brain)
- Endometriosis

This cycle is repeated about once a month, from puberty to menopause—about 400 times in an average woman's lifetime. The average length of the cycle is 29.5 days, but that is only an average. It is quite normal for women to have a cycle ranging anywhere from 15 to 45 days. And, for some women, normal means a cycle length that changes every month. In addition, most women's cycles get shorter as they age, from an average cycle length of 35 days during the teen years to 28 days by the midthirties. Length of menstruation also varies among women, from 2 to 8 days.

A variety of factors can shorten your cycle from month to month, including crash diets, stress, and excessive exercise. (See list on page 108.) But it is the first part of the cycle—or the time before the egg is released from the ovary—that usually is compressed. The time between the release of the egg and the beginning of the menstrual flow is almost always fourteen days. The exception is when the length of the cycle is disrupted by excessive exercise; then, the second half of the cycle appears to shorten.

Why fourteen days? Scientists have no answer, but chronobiologists believe it is somehow tied in with those seven-day rhythms that are so much a part of our biological makeup—and whose source remains a mystery to us.

A TALE OF FOUR HORMONES

Behind the scenes, directing the monthly menstrual cycle—and, to a great extent, how you feel and look—are four sex hormones. Two are produced in the pituitary gland beneath the brain; two in the ovary itself.

Follicle-stimulating hormone (FSH). As its name suggests, this hormone, whose source is the pituitary gland, stimulates ovarian follicles to grow.

Estrogen. This hormone is produced by specialized cells within the developing follicles. Once released into the bloodstream, estrogen builds up the lining of the uterus in anticipation of a fertilized egg. It also thins the mucus in the cervix (the narrow outer end of the uterus) to make it easier for sperm to reach and fertilize the egg.

Luteinizing hormone (LH). Produced in the pituitary gland, this hormone triggers the egg to break from the follicle. It also causes the broken follicle to grow new cells that produce another hormone, progesterone.

Progesterone. This hormone causes the lining of the uterus to secrete protein-rich substances to help nourish the egg, if it is fertilized. It also helps thicken the cervical mucus after the egg has passed through the fallopian tube to make it more difficult for sperm to enter the uterus.

The rise and fall of estrogen and progesterone seem to be responsible for the shifting physical and psychological sensations experienced monthly by many women. When estrogen is in command (generally during the first half of the menstrual cycle), women tend to feel better both physically and emotionally—more self-assured and less irritable. When progesterone is on the rise (generally during the second half of the menstrual cycle), women tend to have more negative feelings, such as low self-esteem, impatience, and lethargy. They also experience more physical ailments, such as headaches, swollen ankles, and muscle fatigue.

These shifts are tendencies that are not experienced by every woman every month. The intensity of the changes also varies from woman to woman—and from month to month for each woman.

The Four Phases of Menstruation

The menstrual cycle, like the moon, has four phases. Here we describe briefly some of the major events of each phase during an average 29.5-day cycle, which, by the way, is also the length of the lunar cycle!

Phase 1: Menstruation (Days 1–5)
  • Estrogen and progesterone reach their monthly lows, causing the lining of the uterus (endometrium) to shed.

- Bleeding begins and may last from two to eight days.
- Selected egg follicles deep within the ovaries double in size.
- Toward the end of this phase, estrogen levels begin to rise.

Mood: Tension of the premenstrual phase (see Phase 4) may go away, but feelings of depression may linger until estrogen levels begin to rise.

Phase 2: Preovulatory (Days 6–14)

- Estrogen continues to rise, reaching peak on day twelve or thirteen.
- Developing follicles move toward the surface of the ovary.
- The endometrium begins to thicken.
- Glands in the cervix create more mucus; during this phase, the mucus becomes thinner and more elastic to make it easier for sperm to enter the uterus.

Mood: Feelings of self-confidence predominate.

Phase 3: Ovulation (Day 14)

- Estrogen levels drop precipitously.
- Immediately after the drop in estrogen, one of the follicles ruptures and releases its egg.
- A sharp, cramplike pain called mittelschmerz (middle pain) may be felt as the egg is released; sometimes bleeding occurs.
- The cervix moves to a position high in the vaginal cavity; the opening to the cervix widens.

Mood: Positive feelings peak.

---

*The moods described represent general tendencies; they are not experienced by all women.

Phase 4: Premenstrual (Days 15–29)

• Within twenty-four to forty-eight hours after ovulation, body temperature rises about one-half to one degree and stays there for several days.

• Cervical mucus becomes thicker and pastier.

• The empty follicle (called a corpus luteum) stops secreting estrogen and begins to secrete progesterone, which reaches its highest levels around day twenty-two.

• The progesterone causes the cells of the uterine lining to secrete protein-rich substances to nourish the released egg.

• Breasts may swell and become tender.

• Endometrium becomes thicker.

Mood: Feelings of anxiety, irritability, helplessness, and depression build.

HOW TO COPE WITH PREMENSTRUAL SYNDROME

Rebecca could always tell when her period was coming. Her breasts would hurt during the first few minutes of her daily run—something that never happened at other times of the month. Her complexion would lose its middle-of-the-month healthy glow and appear duller and more mottled; often a pimple or two would erupt on her forehead or chin. Her mood would also change. She would become anxious about her work and short-tempered with her children and husband. She would also get down on herself during this time of the month, complaining to herself and to her husband that she was physically unattractive and a failure at her job—although, in truth, she was very attractive and had a successful career.

Still, Rebecca came to dread the five or six days that preceded her period. Each month, when those days arrived, all she wanted to do was withdraw from the world and sleep and eat (her two favorite premenstrual activities). Fortunately, a friend recommended a self-awareness course on the menstrual cycle, offered through a local hospital. Rebecca signed up and learned that, by making a few changes in her lifestyle, she could greatly ease her premenstrual blues.

Although the changes experienced by women during their menstrual cycles can be positive as well as negative (see Cycles Change list on page 120), most attention seems to have been focused on the more undesirable psychological and physical symptoms associated with the final, premenstrual phase of the cycle. Known collectively as premenstrual syndrome, or PMS, these symptoms include such things as headaches, fatigue, depression, anxiety, and irritability. In the past, PMS has been cited as evidence that women are incapable of holding important jobs or public office. It should be pointed out, however, that only about 5 to 10 percent of menstruating women suffer symptoms severe enough to cause them to miss work or be unable to perform their regular work assignments. The vast majority of women cope competently with premenstrual complaints. Some women actually welcome the feelings of enhanced creativity that often precede their periods and have learned how to channel those feelings into highly creative and productive work.

Many women, however, would like to change how they feel before their periods—even if it's a relatively mild change, such as feeling less bloated or tired. If you are one of these women, there are several actions you can take to minimize or eliminate the discomfort of PMS.

• *Watch what you eat during the two weeks prior to your period.* Studies suggest that changes in your hormones make it more difficult for your body to process carbohydrates right before your period. So, try to avoid sugary junk foods, since your body is more likely at this time of the month to overreact to the sugar and clear too much of it from the blood. The result: low blood sugar and physical symptoms such as headaches, dizziness, and shakiness.

Cutting down on junk food isn't always easy, since a craving for sweets is, for many women, one of the most persistent premenstrual symptoms. Try stocking up on fresh fruit and getting rid of any sweets or biscuits you may have around when you know your period is on its way. Then, when the craving hits, you won't be so easily tempted. Eating smaller, but more frequent, meals during the premenstrual phase can also help lessen sugar cravings by keeping your blood sugar level steadier. For a lean and healthy body, however, be sure that those meals are low in fat.

• *Cut back on salt.* It increases the body's ability to retain fluid. During the premenstrual phase, the body has more difficulty eliminating fluid. Scientists are not sure why this happens, but some believe it is related to the drop in progesterone that occurs immediately before menstruation begins. Not only does excess fluid cause premenstrual swelling symptoms, such as weight gain, breast tenderness, and abdominal bloating, but the fluid also gathers in brain tissue, resulting in mood swings.

Cutting back on salt means more than just tossing away the salt shaker. Most of the salt we eat is hidden in processed foods, including cheese, sandwich meats, canned vegetables, soups, and breakfast cereals. The food we buy at fast food restaurants is also heavily salted. Read food labels carefully; look at the amount of sodium listed.

• *Increase the amount of water you drink.* Water is an excellent diuretic and will help rid your body of excess fluid.

• *Take selected vitamins.* Some women find daily dosages of calcium (500 mg), magnesium (250 mg), and vitamin B6 (less than 250 mg) helpful in reducing the emotional symptoms of PMS, such as depression, irritability, and anxiety. *A caveat:* Always take B6 as part of a B-complex vitamin, because too much of one B vitamin can cause a depletion in the body of the others.

• *Avoid alcohol.* During the premenstrual phase, many women have a lower tolerance of alcohol. One glass of wine may have the same effect as three glasses at another time of the month. Also, because alcohol is a depressant, it may intensify the negative thoughts or feelings that are so often a part of PMS.

Scientists are not sure why women have a lower tolerance of alcohol before their periods, but they believe it may be tied to the monthly rhythms of estrogen.

• *Avoid caffeine.* Like alcohol, caffeine can aggravate many of the emotional symptoms of PMS. Be aware that caffeine is found in chocolate, soft drinks, and many over-the-counter medications, as well as in tea and coffee. (For a list of over-the-counter medications containing caffeine, see page 184.)

• *Do not take tranquilizers.* They can intensify the depression and other psychological symptoms that are part of PMS.

• *In general, take care of yourself.* Listen to your body. If you need more sleep, get it. If you feel less inclined to be with people, excuse yourself from social commitments and arrange for time alone. Rest, relax, and be good to yourself! And remember: This, too, shall pass.

IF SELF-HELP MEASURES DON'T HELP

Sometimes, self-help measures don't work for women who suffer from more disruptive forms of PMS. If this is true for you, you should discuss medical treatments for PMS with your physician. These treatments include progesterone injections, birth control pills, and antidepressant medications. All have side effects, so you should attempt such treatment only under the supervision of your physician.

When Your Period Is Most Likely to Start

Your menstrual period is most likely to begin in the morning, between 4:00 A.M. and noon. It is least likely to start in the evening. This may have to do with the daily rise and fall of hormones, or it may simply be that you tend to notice any period that starts during the night only after you get up in the morning.

Before you begin treatment, however, you should be aware that no treatment has been found to be the definitive cure for PMS. Progesterone treatments are particularly controversial; recent research has cast doubt on their effectiveness. Indeed, in a 1979 British study of the effectiveness of progesterone on PMS—one of the few studies that was double-blind, or conducted so that neither the researchers nor the subjects knew which treatment was which—a placebo was shown to be just as effective as progesterone in relieving premenstrual symptoms.

CHARTING YOUR MONTHLY CYCLE

Each woman has a unique rhythm to her menstrual cycle. By charting your cycle for three or four months, you will begin to recognize the highs and lows of your monthly rhythm—the days when you feel like going out and conquering the world and the days when you prefer to curl up in front of the fire with only a good book for company.

Charting is a simple process and requires only a small amount of time. Here's how to go about it:

• Use either a monthly calendar with large daily squares or make a calendar using large pieces of paper. (See example on page 119.)

• Select ten physical or psychological changes that you wish to keep track of from the list on page 120 and write them on the calendar page. Give each one a letter symbol: I for irritability; BT for breast tenderness, etc. Try to include at leat two positive changes.

• Each day, record any noticeable changes on your chart. If you miss a day, leave the square blank; don't go back and try to fill in blank spaces. If you don't notice any changes on a particular day, simply write "no change."

• If a symptom is very strong on a particular day, circle the letter symbol on the calendar.

• Use a star to indicate the first day of your period and, if you wish, another star to indicate when the period stops. During your

What One Woman Learned from Her Chart

About halfway through her twenties, Joan began to suffer fairly strong PMS symptoms—tension, fatigue, sleepiness, anxiety, and, most worrisome to her, deep depression. She recognized almost at once that the symptoms were related to her monthly cycle and decided to look for ways she could change her lifestyle to make the week before her period more bearable.

So she began to chart her cycle. She also began experimenting with changes in her diet to see if food could ease the symptoms. Nothing seemed to work. Then, after almost a year and a half of charting, she noticed a pattern: The symptoms were worse during the winter months.

Joan lived in a northern city where sunlight was scarce in the winter. Maybe, she told herself, the lack of light was contributing to her PMS. As an experiment, she started going to a tanning booth for just fifteen minutes a week. Within two months, her symptoms had all but disappeared! She also noticed that her periods were more regular.

Joan's self-treatment might not work for everyone. Indeed, many health care specialists warn about the serious health consequences of overexposure to ultraviolet rays. Also, Joan's PMS symptoms seemed to be compounded by seasonal depression (see chapter 4)—a problem most other PMS sufferers do not share. But her efforts at tracking her cycle illustrate how charting and finding patterns in your cycle can help you take action against the PMS blues.

period, you may also wish to indicate the pattern of your bleeding: heavy, moderate, or light.

• You may also keep track of your weight during the month by writing it down in each daily block. Be sure to weigh yourself at the same time each day.

Keep charting for at least three months. The longer you chart, the clearer will be the pattern that emerges. Many women find that knowing a pattern exists is enough to help them live more comfortably with their monthly ups and downs. They realize that their symptoms are not "in their heads" at all, but a natural part of their monthly cycle.

MEN HAVE CYCLES, TOO

Although little research has been done on the topic, men have monthly cycles, too. As far back as the seventeenth century, the Italian scientist Sanctorius weighed men daily over long periods of time and discovered that men underwent a monthly weight change of about one or two pounds. More recently, a Danish endocrinologist kept daily records of hormones excreted in his urine. When analyzed, those records showed that his hormones rose and fell in roughly a thirty-day rhythm. Interestingly, beard growth also shows a rhythm of approximately thirty days; in other words, the amount of beard a man grows daily increases and decreases in a monthly cycle.

In 1929, a researcher carefully followed the moods of seventeen men and showed that men, like women, have emotional cycles of about a month to six weeks in length. According to the researcher's findings, men tend to be more apathetic and indifferent during the low period of their emotional cycles and more likely to magnify small problems into big ones. During the high period of their cycles, men have more energy, a greater sense of well-being, lower body weight, and less need for sleep.

In the early 1970s, *Ms.* magazine reported that a Japanese bus and taxi company used this knowledge of male monthly cycles to reduce accidents involving its vehicles. After charting the cycles of

A Monthly Record

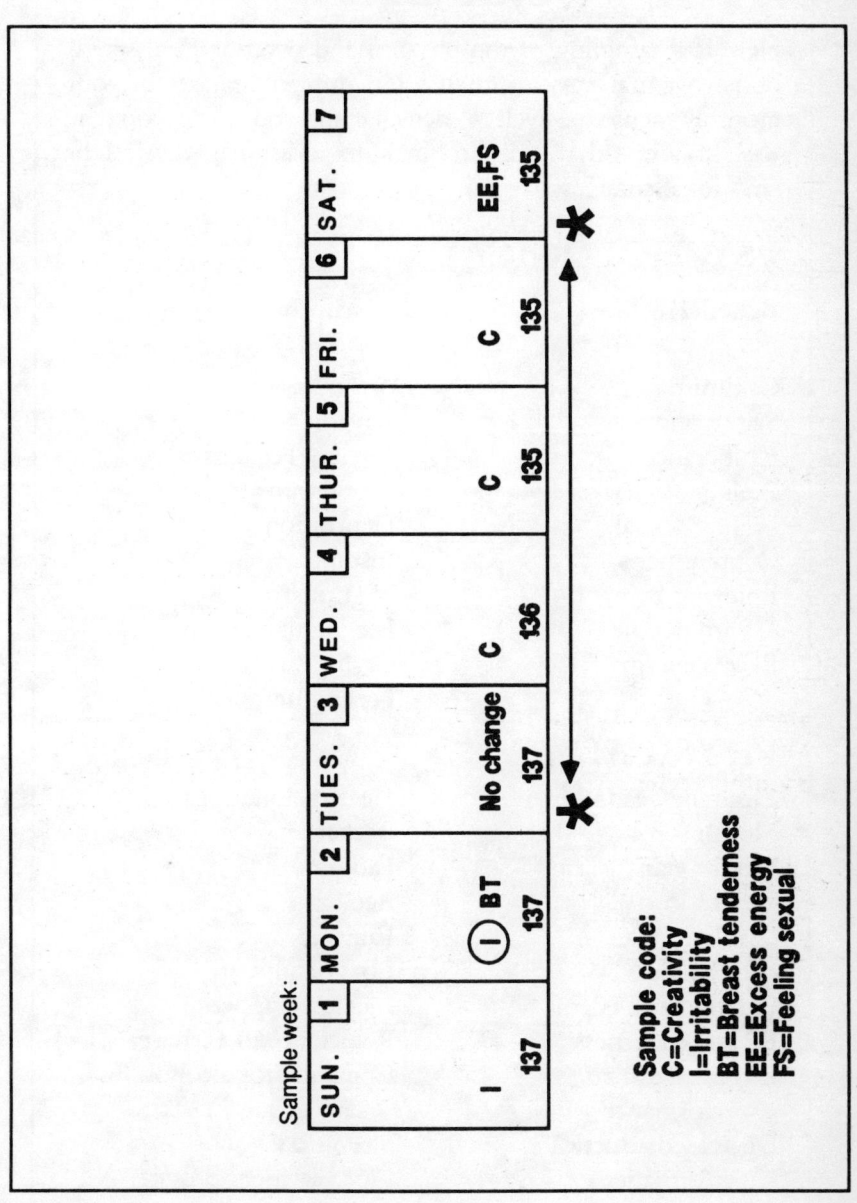

Cycle Changes

Here is a sampling of the physical and psychological sensations that may vary in intensity at different phases of your monthly menstrual cycle. When charting your cycle, you may wish to expand this list to include additional physical or emotional changes:

PSYCHOLOGICAL

Beneficial

Creativity
Optimism
Self-confidence
Exuberance
Feeling attractive
Feeling sexual
Contentment
Enjoying being
  with people
Excess energy

Disruptive

Irritability
Anger
Tension
Anxiety/Feelings
  of doom
Depression
Insomnia
Nightmares
Feeling unattractive
Crying spells
Feeling unsocial

PHYSICAL

Migraine headaches
Mouth sores
Weight gain
Weight loss
Swollen ankles
  or hands
Bloating and edema
Breast tenderness
Increased athletic
  performance
Decreased athletic
  performance

Heart palpitations
Backache
Fatigue
Acne
Fainting
Loss of appetite
Increased appetite
Reluctance to exercise
Cravings (for alcohol,
  salt, or sugar)
Sleepiness
More accident prone

each of its drivers, the company rearranged routes and schedules to best match the men's low and high periods. The result: After two years, the company experienced a one-third drop in its accident rate.

Men, like women, may benefit from following the Japanese company's example and charting their individual monthly rhythms. Use the chart and Cycle Changes list on page 120. If you are married or living with a woman, both of you may find it interesting and helpful to chart your cycles together to see if they are in or out of sync with each other.

SHORTER MALE CYCLES

In addition to its monthly rhythm, beard growth has a second—and weaker—weekly rhythm in men. Beards tend to grow the most on Sunday and the least on Wednesday—although the difference is not great enough for most men to notice. Because beard growth seems to be triggered by testosterone levels (which rise during sexual activity), this weekly increase and decrease in beard growth may be the result of increased sexual activity on the weekend.

One of the most perplexing and shorter male cycles involves the rise and fall in sperm count. A man's total sperm count peaks every two to five days; for most men, the peak comes every three or four days (which is also, by the way, the same rhythm exhibited by male rabbits). What causes the sperm count to rise and fall in such a rhythm or how it affects sexuality is still being explored.

HOW YOUR MENSTRUAL CYCLE AFFECTS YOUR PHYSICAL HEALTH

Your susceptibility to illness changes throughout your monthly cycle. For example, a woman in the middle third of her cycle is dramatically more susceptible to colds. In one study, 77 percent of women exposed to cold viruses during the middle of their cycles developed colds, compared to 29 percent of other women. Estrogen may be the cause. It thins the cervical mucus in the middle of the cycle to encourage sperm to reach the ovulating egg. In the process, it may also be thinning the mucus produced by the membranes of the nose, and thus be giving cold germs easier access to the body. Cold germs usually invade the body through either the nose or eyes.

When the cold and flu seasons hit, therefore, you should probably take extra preventive precautions during the middle of your menstrual cycle, when you are most susceptible. Try to avoid touching your face with your hands. Wash your hands frequently, especially if you've just been in the same room with someone who has a cold. Also, avoid using the cold sufferer's towels, washcloths, and face soaps.

Despite the increased susceptibility to colds at the time of ovulation, a woman's immunity to other ailments is usually strongest during this phase of the cycle—and at its weakest right before and during menstruation. So, the second half of a woman's cycle is when other ailments are most likely to strike.

Asthma attacks, for example, are more severe immediately before and during the menstrual period. Scientists believe the severity of the attacks may be the result of the fluid that builds up in the body at that time, which causes the lining of the airways to swell, making it more difficult to breathe. Doctors recommend that you reduce your salt intake before your periods to keep fluid buildup to a minimum. You may also want to consult your physician about increasing your medication at this time of each month.

Herpes sores are also more likely to appear around menstruation. In fact, the timing of the sores during menstruation is so common that one clinical form of the disease is called *herpes menstrualis*.

Getting ill isn't the worst thing that is more likely to happen to you during the second half of your cycle. You have a greater chance of dying then, too! In one study, researchers looked at the postmortem coroners' reports of 102 women, age eighteen to forty-six, who had died of various causes. Much to the researchers' amazement, they discovered that only 13 of the women had died during the first half of their cycles; the other 89 died between ovulation and menstruation.

A closer look at the coroners' reports showed an even more startling fact: 60 of the 89 women who had died during the second half of their cycles did so during days seventeen to twenty-three.

When an Illness Is Most Likely to Strike

The following medical illnesses and conditions are most likely to occur during a particular phase of the monthly menstrual cycle:

Ovulation	Premenstruation	Menstruation
Colds and flu	Nose bleeds	Hives
	Hoarseness	Herpes
	Headaches	Pharyngitis
	Acne	Edema
	Eczema	Asthma attacks
	Tetany (muscle spasms)	Peptic ulcer attacks
	Asthma attacks	Tonsillitis
	Pneumonia	Diabetic coma
	Epileptic attacks	
	Inflammation of the pancreas	
	Hepatitis	
	Scarlet fever	
	Typhoid fever	

Granted, the study cannot be considered representative of all deaths, because coroners usually deal with suicide and accident victims—and several studies indicate that suicides and accidents are more prevalent during the premenstrual phase, when women tend to be more depressed. Indeed, 50 percent of the deaths in this study were due to those causes; the figure for the general population is closer to 5 percent.

Still, the fact remains that nearly *all* the deaths—not just suicides and accidents—occurred during the last half of the menstrual cycle. This is a startling finding, and one that needs to be explored further. It may be that one day women will be advised to schedule elective surgery during the first half of their monthly cycles.

HOW YOUR MENSTRUAL CYCLE AFFECTS YOUR INTELLECTUAL ABILITIES

No one knows just how—or if—a woman's intellectual abilities are affected by her monthly menstrual cycle. Some studies have shown that a woman's reaction time is slowest just prior to menstruation; other studies have refuted this finding. A few studies have also shown that women score lowest on intelligence tests during menstruation; but those results may reflect the physical discomfort of menstruation (known as dysmenorrhea) rather than a true change in intellectual abilities.

Most studies seem to indicate that a woman's intellectual skills are *not* significantly affected by the hormonal fluctuations of her menstrual cycle.

Still, it may be wise to try, if possible, to schedule any major job- or school-related tests on nonmenstruating days. You should do this for two reasons. First: Menstrual aches and pains may keep you from performing your best. Second: Studies have shown that some women perceive themselves as doing poorly on tests during this time of the month—and such perceptions can become self-fulfilling prophecies.

Of course, most tests can't be scheduled at your convenience. If you must take a test or perform some other intellectual feat during menstruation, don't panic. Rather, take these simple steps to ease the discomfort of menstruation:

• *Get some exercise in the days preceding the test.* Exercise will help rid the body of excess water, ease the flow of blood, and relieve the constipation that often accompanies menstruation.

• *Get enough rest.* Often, women who are menstruating need extra sleep.

• *Take aspirin for painful cramps* (assuming you are not allergic to aspirin). If the cramps are particularly severe, ask your doctor about several anticramp prescription drugs now on the market. You should also have the doctor examine you to make sure the cramps are not the result of a more serious underlying condition, such as endometriosis (an abnormal growth of uterine tissue) or pelvic inflammatory disease (PID).

HOW YOUR MENSTRUAL CYCLE AFFECTS YOUR ATHLETIC ABILITIES

If you were to enter an athletic event during your menstrual period, do you think your performance in the event would be better or worse than it would be at some other time of the month?

Was your answer "worse"? Well, think again. Although you, like many other women, may believe you perform more poorly at sports or athletics during your period, you may actually perform better.

Here's what some of the studies show:

• When female competitors at the 1930 World Track and Field Championships were asked how their menstrual cycle affected their performance, *29 percent said their performance improved while they were menstruating*.

• Twenty-two years later, at the 1952 Helsinki Olympics, another group of top female athletes was asked the same question.

How Exercise Can Affect Your Monthly Cycle

Many female athletes report that their menstrual periods stop or become infrequent when their training gets into high gear. The reason? Intense exercise over a long period of time appears to alter the balance of hormones within the body in such a way that ovulation seldom or never occurs. And, of course, without ovulation, menstruation does not occur.

Some female athletes are more likely to have their periods stop than others. Here's a look at the high-risk groups:

- Very lean women—specifically those with less than 20 percent body fat
- Women who weigh less than 115 pounds
- Women who lose more than 15 percent of their total body weight after beginning an exercise program
- Women who regularly consume fewer than 1,500 calories per day
- Women who are vegetarians—probably because they consume much less fat in their diets than nonvegetarians

It's not known what effect the absence of menstruation (known as *amenorrhea*) has on a female athlete's long-range health. In the short term, of course, it causes infertility.

Amenorrhea is reversible, often by simply gaining a few pounds or lowering the number of miles you run, swim, or cycle each week. Always consult first with your doctor to make sure the change in your cycle does not have a more serious cause. Also, be sure to apprise your doctor that you are physically active; otherwise, he or she may overlook this as a cause of amenorrhea, and you may find yourself receiving unnecessary hormone treatments.

Again, *20 percent reported an improved performance during menstruation*.

• More recently, researchers tested a group of competitive teen-age swimmers through several menstrual cycles. *All recorded their fastest practice times during menstruation.*

Some studies do indicate a deterioration of strength, hand steadiness, and balance around the menstrual period—particularly on the first day of flow. But just how much of an effect this has on overall sports performance is not clear.

The best way to keep your athletic performance steady throughout the month is to keep in shape. Women who exercise regularly complain less about their periods than those who do not. But, it's not for the reasons you think. Although exercise can shorten your periods and decrease the flow of blood, it does not appear to lessen the discomfort of dysmenorrhea—or, for that matter, of PMS. A recent survey of 420 women by researchers at the Melpomene Institute for Women's Health Research revealed that exercise has no effect on menstrual cramps, mood swings, or monthly changes in appetite.

Why, then, do women who are physically active complain less about their periods? No one knows for sure, but it may be simply that they are more used to discomfort and thus better able to tolerate it. It also may be that, because these women tend to be in excellent health, their bodies are better able to handle the strain—and occasional pain—of menstruation.

HOW YOUR MENSTRUAL CYCLE AFFECTS YOUR WEIGHT

As many women know, stepping on the bathroom scale during the days preceding menstruation can be a depressing experience. Most women will register a weight gain of three to five pounds during the premenstrual phase of their cycle. It's important to remember, however, that this weight gain is only temporary, caused by the excess fluid retention so common during that time of the cycle. The extra weight is usually quickly shed once menstruation begins.

Of course, you can gain permanent weight during the premenstrual phase if you give in to cravings you may have for sugary or fatty foods. And many women do give in to those cravings, consuming an average of 500 extra calories each day during the ten days before their periods! Fortunately, your basal metabolic rate (the minimum amount of energy you need to keep your body working) is at its highest right before menstruation, and a higher metabolic rate means that you burn calories more easily. That, however, does not mean that you can wolf down a pint of chocolate ice cream and not expect it to find its way to your thighs. We're talking about burning only a *very* few extra calories!

If you are thinking about going on a diet, the best time to do it is right after your period. Many women find that their desire for sweets is noticeably reduced during the preovulatory phase of their menstrual cycle. Obviously, the worst time to start a diet is right before your period, when your craving for food will be strongest. In fact, you are five times more likely to quit your diet at this time of your cycle.

HOW YOUR MENSTRUAL CYCLE AFFECTS YOUR SKIN

Premenstrual flare-ups of acne occur in half of women who are already prone to breaking out. Experts believe these flare-ups are linked to the premenstrual rise of progesterone, which activates the oil-producing glands of the skin and causes blackheads and pimples to form. On the other hand, estrogen, which dominates the first half of the menstrual cycle, inhibits oil production. Your skin will generally look its clearest, therefore, just prior to ovulation. This may also account for the clear-skinned glow of pregnant women, for they have higher levels of estrogen in their system than nonpregnant women. Many women taking high-estrogen oral contraceptives also report that their skin clears up once they are on the Pill.

You can minimize premenstrual acne by taking special care of your skin during the two weeks prior to your period. Wash your face with soap at least three times a day. Pat it dry gently with a soft towel; don't rub or scrub. Remove excess oil with cotton swabs dabbed in alcohol or apply a lotion containing benzoyl peroxide, an antibacterial agent that dries and protects the skin against excess oiliness. Use a water-based or oil-free makeup foundation instead of an oil-based one that can clog pores.

During the first two days of your period, you should also take extra precautions to protect your skin from the sun, since at this time of your cycle, your skin is especially sensitive to the sun's

---

A Special Note to Women on the Pill

Women who take birth control pills have less pronounced monthly ups and downs than women who use other methods of birth control. This is because birth control pills prevent ovulation, usually by providing a steady dose of estrogen that is too high to allow an egg to mature. If you are on the Pill, you can probably expect shorter and lighter periods, fewer problems with acne, and less severe shifts in mood.

Of course, you may experience other discomforting symptoms, such as breast tenderness and weight gain (all month, not just before menstruation), skin rashes, and an increase in vaginal discharge. Birth control pills have also been linked to high blood pressure, depression, skin cancer, and other illnesses. These complications are rare, however.

Still, the Pill is a good example of how changing a natural rhythm can create serious health consequences—and why we need to treat our rhythms with respect.

ultraviolet rays and can burn more easily. Either stay out of direct sunlight (especially if you have fair skin) or use a sunscreen lotion or cream, preferably one that contains the chemical PABA or benzophenone, or both.

HOW YOUR MENSTRUAL CYCLE AFFECTS YOUR SEXUAL DESIRES

Women usually feel their sexiest at midcycle, or right around ovulation. This makes sense from a biological perspective: Having women most receptive to sex at the time they are most likely to become pregnant helps ensure continuation of the species.

Scientists are not sure what causes this increased sexual libido at midcycle, but they believe it may be related to a monthly peaking of androgens.

According to one study, women increase their sexual desire and activity by 25 percent during a three-day period around ovulation. This is not true, however, of women who are taking birth control pills and who, therefore, are not ovulating. In fact, during the time they would normally ovulate, they actually experience a decreased desire for sex, although researchers are not sure why.

Not only is a woman more likely to want sex at ovulation, but she is also likely to enjoy it more as well. One study revealed that women are two to six times more likely to have an orgasm at midcycle. Contributing to this heightened enjoyment of sex is the fact that a woman's senses—taste, touch, smell, and hearing—are more acute and sensitive at ovulation. Interestingly, a woman's skin is more sensitive to touch at this time of the month (although less sensitive to pain), further encouraging the woman's receptivity to sex.

A woman's sense of smell is particularly strong at ovulation. In fact, only women who are ovulating are able to smell certain artificial musklike odors—odors that may mimic the natural odors, or pheromones, emitted by men. This ability to smell these odors only during ovulation may play a role in triggering a woman's increased sex drive.

Some sex researchers have reported that women have a second stretch of days during their monthly cycle when they tend to be more sexually active—right before menstruation. This peak may be caused by the sudden drop in progesterone that immediately precedes menstruation. It is believed that high levels of progesterone suppress female libido; when the levels fall, therefore, desire rises.

HOW BEING AROUND WOMEN AFFECTS YOUR CYCLE

As sisters and roommates have often discovered, living with another woman or group of women can eventually result in everyone menstruating at around the same time of the month. This has been called the "dormitory effect," because the first formal study of the phenomenon was conducted among a group of college students living in a dormitory. Generally, it seems to take about three or four months for menstrual cycles to become synchronized. For example, the cycles of seven female lifeguards involved in one study were far apart when the women began working together at the start of the summer, but fell within four days of each other three months later.

Scientists are not sure what causes menstrual synchrony, but they believe it may be a female pheromone that in some unknown way triggers changes in the menstrual cycles of other women. At least one study backs up this theory. In the study, perspiration was taken from one donor woman and rubbed on the upper lips of five other women. Six control women had their lips rubbed with plain alcohol. The menstrual cycles of all the women were then followed. The cycles of those who received the perspiration shifted toward the donor's menstrual cycle. The control group showed no such shift.

HOW BEING AROUND MEN AFFECTS YOUR CYCLE

Some research suggests that if you are regularly having sex with a man (at least once a week), your menstrual cycle will tend to be shorter and more regular than it would be if you were celibate.

Again, pheromones—this time, male ones—may be the cause, since research has revealed that women who work in the manufacturing of natural and synthetic musks—substances related to testosterone, the most dominant sex hormone in males—also have shorter and more regular cycles than other women.

Tips for Living with Your Sexual Rhythms

Sexual rhythms can be subtle. Here is a summary of tips for ensuring that these rhythms enhance rather than interfere with your life.

• Chart your monthly cycles. Be aware of your peaks and valleys in terms of physical and emotional well-being, and try to schedule your activities accordingly.

• As noted, you may feel sexier in the morning. Take advantage of this possible increase in sexual libido by scheduling more lovemaking with your partner before you get out of bed in the morning.

• Be cautious if you fall in love in late summer or autumn. The irresistible attraction you feel at that time of the year may not be love at all, but rather a seasonal peaking of sexual desire.

• For women only: If you do not want to become pregnant and you are not using birth control or are using natural family planning, be aware that the time of the month when you most desire sex may very well be the time when you are most likely to become pregnant.

Six

Keeping a Healthy Beat—
Rhythms and Medicine

Healing is a matter of time . . .
—HIPPOCRATES
Precepts

Kay was a wreck. It was a week before Thanksgiving and her asthma, which had hit with a vengeance in mid-August and been getting steadily worse, was now so bad she could hardly stand it.

The same thing had happened last Thanksgiving, except then it had culminated in an early morning attack so severe that her frantic roommate had called the emergency services. Within minutes, an ambulance had arrived and rushed Kay to the hospital. The doctors later told her that the attack could have killed her. So this year she had been extra careful about taking her medicine. As a result, the attacks weren't as bad, but they still were difficult.

Kay thought back to the sudden asthma attacks of her child-hood, attacks that often meant a few days in a hospital bed under a clear plastic oxygen tent. She remembered the moist, easy-to-breathe air that was cooled by ice packed in a container near her head. She remembered being frightened by the distorted images of her parents as she looked at them through the soft plastic walls of the tent.

Then it struck her. She had almost always been hospitalized in the autumn. Kay had long been aware that the attacks usually came late at night or early in the morning, but she'd never thought of her asthma as having a seasonal rhythm as well.

Then another thought clicked in her mind. She pulled out a calendar and a note pad and started marking down the dates of her most recent and worst bouts of wheezing and breathing problems. When she was finished, she noticed yet another pattern. The attacks seemed to be more frequent and intense just before and during her menstrual period.

The daily, monthly, and yearly patterns that Kay discovered with her asthma are not unique either to her or to her illness. Just as the human body is not static in health, neither is it static in sickness.

In fact, it is in the prevention, detection, and treatment of disease that chronobiology holds the greatest promise for society.

Already, some doctors are successfully using their knowledge of biological rhythms to time the administration of medicines, such as the anticancer drug Cisplatin, so that they do the most good and the least harm to the body. In the future, biological rhythms

---

Your Susceptibility to Illness:
A Quiz

True or false?

- You are less likely to have a headache on Monday than on any other day of the week.
- If you suffer from asthma, you are more likely to have an attack of the illness at night than during the day.
- Your chance of suffering a heart attack is greatest in the morning.
- You are more likely to find a cancerous tumor in your breast in April than in September.
- Your chances of catching gonorrhea are greater during August and September than at any other time of the year.

All of the answers to the above are true.

may also be routinely used to predict—and help prevent—such illnesses as heart attacks and cancer.

In this chapter, we'll examine the role rhythms play in your physical health. First, we will take a general look at how an understanding of biological rhythms may be used in the future to prevent, detect, and treat disease. Then we will look in more depth at the cycles of some specific illnesses—and how you can use a knowledge of these cycles to your health advantage today.

USING RHYTHMS TO DETECT DISEASE

A sleepless night, a missed menstrual period, a sudden, high body temperature—all are biological rhythms gone awry and all are possible signs of an underlying illness. For example:

• A change in the sleep cycle can be an indication of clinical depression.

• A change in the menstrual cycle can be an early sign of endometriosis or thyroid disease.

• And, of course, a change in body temperature can indicate one of a host of illnesses, from the common cold to meningitis.

You should be alert, then, to these and other changes in your rhythms, and report any that are prolonged or severe to your doctor.

Significant changes in our sleep, menstrual, or temperature cycles are usually obvious, even to those of us who pay little attention to our bodies. Recently, however, chronobiologists have begun to study more subtle rhythms to see if they, too, will be useful in the early diagnosis of disease. What they have discovered is promising.

For example:

• *Breast cancer*. Healthy breasts have regular daily, weekly, and monthly surface temperature rhythms. Cancerous breasts—even those with small, hidden tumors—do not. When cancer strikes a breast, its daily surface temperature shifts from a normal twenty-four-hour rhythm to a shorter, ultradian rhythm of about twenty hours.

One day soon, women may be regularly monitoring the temperature of their breasts for early signs of cancer—just as they now give themselves monthly breast self-examinations. Chronobiologists have already developed prototypes of temperature-monitoring bras, and they hope that simplified, less expensive versions will eventually become available.

These bras will be especially helpful to women who are in the high-risk category for breast cancer—women, for example, who

---

Your Health Thermometer

If you wake up in the morning with a temperature of 99° F, you are very likely coming down with a cold, the flu, or some other kind of illness. But if you have that same reading around dinner time, you may be in perfect health. The reason, of course, is that your temperature rises during the day. A high reading *early* in the day is a greater indication of illness than the same reading late in the day.

You can be even more specific than that, however, in deciding from your temperature whether or not you are ill. All you need to do is record your temperature at different times of a typical healthy day as suggested in chapter 2, and then use those readings as a control against which to judge readings taken on other days.

This can be a great diagnostic tool on those mornings when you wake up feeling a bit out of sorts, but can't decide whether you are coming down with something—and therefore whether you should stay home from work. Parents can also use it to help decide whether a child is well enough to go to school or day care.

have close blood relatives with the disease. The bras could be worn for a week every several months to check for telltale changes in temperature rhythms.

• *Arrhythmia (irregular heartbeat)*. People who suffer from arrhythmia, or an irregular heartbeat, may also benefit one day from the use of diagnostic rhythms. Arrhythmia is a potentially deadly condition, for an erratic heartbeat can sometimes cause the heart to stop beating altogether.

After a five-year study, scientists discovered that the incidence of sudden death in arrhythmia patients is much higher among people who show a particular pattern and timing to their arrythmia. In other words, the patients who died had heartbeats that formed a regular *irregular* pattern.

Unfortunately, this dangerous pattern was noticed only after extensive electrocardiograph monitoring. Such monitoring is not yet feasible for all arrhythmia patients.

As you can see from these two examples, the impracticality of tracking rhythms outside of carefully controlled studies is one of the things that is delaying the use of rhythms in diagnosing diseases. But, as monitoring equipment improves and doctors become more aware of the diagnostic capability of our rhythms, detecting serious illnesses long before they have revealed any outward symptoms may become commonplace.

USING RHYTHMS TO TREAT DISEASE

When your doctor prescribes medication for you, he or she most likely tells you to take the medication at arbitrary intervals, such as "four times a day" or "every six hours" or "with meals." The idea is to make the taking of the drug simple to remember and to ensure that the drug is constantly in the body and doing its job.

But chronobiologists have discovered a more effective way of prescribing medications—one based on precise timing as well as dosage. They have found that *when* you take a therapeutic drug may be as important as—or perhaps even more important than—how much of the drug you take. In other words, the potency of most drugs depends dramatically on when they are administered.

The reason? As our bodies change throughout their twenty-four-hour cycles, they become more or less susceptible to various drugs. The same amount of a particular drug may be *much more effective* if administered at, say, 9:00 A.M. than at 9:00 P.M.

Take a commonplace drug like aspirin, for example. When swallowed at 7:00 A.M., aspirin stays in the body for up to twenty-two hours. When the same dose is taken at 7:00 P.M., it is completely out of the body within seventeen hours.

Or, look at a more specialized drug, such as the anticancer drug Cisplatin. This drug is extremely powerful and can severely and permanently damage the kidneys—one of the risks of using it for chemotherapy. However, chronobiologists have discovered that if Cisplatin is administered in late afternoon, when the kidneys are most active and secreting the most potassium, the drug does the least amount of damage to the kidneys. If it is taken late at night when the kidneys are least active, it remains in the kidneys for a much longer period and does more damage.

The implications of these findings are enormous. Powerful drugs can be administered at those times of day when they do the most good and have the least side effects.

Already, some pharmaceutical companies are considering redesigning drug labels with daily rhythms in mind. Soon consumers may find that a drug label that said "take once a day" has been rewritten to say "take at 3:00 P.M." Old habits are hard to break, however, and it's not clear how quickly consumers—or, perhaps even more importantly, their doctors—will accept the change.

Drug therapy isn't the only medical treatment that benefits from knowledge of biological rhythms. Radiation therapy—the use of high energy rays to destroy or alter cancer cells so that they no longer reproduce—also varies in effectiveness depending on the circadian rhythms of patients.

Perhaps the most dramatic example of this comes from an experiment done in India involving a group of people who had a

Pumps That Keep the Beat

Administering drugs in a timely manner can be lifesaving, but it can also be inconvenient for doctors, hospitals, and patients. Drug therapy can stretch out for many weeks, months, or even years, and often more than one kind of medication is involved. Stopping to take the medication at precise times may not always be practical—especially if one or more of the drugs must be administered in the middle of the night. It is also nearly impossible for a patient to self-administer at home a low but continuous dose of medicine over several hours—a drug schedule required by some chemotherapy treatments.

To make drug timing more practical, chronobiologists in the US have begun using a sophisticated portable drug pump, developed by a small company in Colorado, that allows up to four drugs to be administered on a preprogrammed basis. At the University of Minnesota Hospital, for example, the pump is being used by chronobiologists to administer complex chemotherapy treatments for cancer patients. The pump also allows patients who otherwise would have to stay in the hospital for complicated treatments to receive their treatments at home. The pump was approved by the US Food and Drug Administration in 1986.

Other drug pumps, some of which are small enough to be implanted into a patient's chest, are also being used to provide timed administration of drugs. But these smaller pumps generally only hold one drug at a time and are limited to simple treatment schedules.

penchant for chewing a mixture of betel nut, tobacco, and lime. The habit often resulted in a rare form of cancer in which a large tumor formed in the mouth. Fortunately, the location and size of the tumor made it easy to isolate and treat with radiation.

Doctors treating these unusual tumors discovered that the tumors had a daily temperature cycle that peaked at a time different from that of the healthy surrounding tissue. To determine if this temperature rhythm had any effect on the success of the radiation treatment, they decided to conduct an experiment. Some tumors were irradiated at the time of peak temperature; others were irradiated eight hours later.

The results were striking. Tumors treated when they were hottest *shrank by 70 percent* within five weeks. Those treated eight hours after the temperature peak, shrank by only 30 percent.

Even more importantly: Two years after treatment, only 13 percent of the patients who had been treated at the nonpeak time were cancer-free. *But 60 percent of the other group showed no further signs of cancer!*

Although the radiation treatment helped both groups, it was clear that the timing of the treatment made a dramatic—and, in some cases, lifesaving—difference.

USING RHYTHMS TO PREVENT DISEASE

Knowledge of your individual rhythms may one day help you extend your life by helping you predict just how susceptible you are to what chronobiologist Franz Halberg calls the "handicapping diseases"—diseases for which some of us, because of our genetic history, are at a greater risk. These include high blood pressure, heart disease, stroke, cancer, and certain diseases of the kidney.

Once you know you are in a high-risk category for a particular illness, you can then take the necessary preventive steps, such as frequent medical checkups or a special diet or exercise program, to keep the illness at bay.

Falling Ill When You're Out of Sync

It never failed. Whenever Linda went on her annual big trip vacation, she would always fall ill. It didn't matter where she went—Japan, Barbados, Hawaii—she always seemed to come down with a cold or flu virus shortly after arriving at her long-awaited destination.

"Why me?" she would ask herself as she lay curled up in her hotel bed with aspirin for her fever and tissues for her running nose. "And why *now?*"

What Linda experienced on her vacations is not uncommon. Many people travel to exotic—and expensive—places only to find themselves spending part of their time at the new destination nursing a cold or a flu.

Why? The answer has a great deal to do with stress and biological rhythms. When you travel, you throw your rhythms out of sync and your body into a stressful state, and that can make you more susceptible to illness.

Of course, you don't have to travel across time zones to have your rhythms desynchronize. Even at home, a stressful situation that breaks up your routine—such as working late night after night on an important project or getting up in the middle of a night to care for a newborn infant—can throw your rhythms out of whack and make it harder to fend off a viral or bacterial infection.

What you can do. First, when you know your rhythms are out of sync, be aware that you are particularly vulnerable to illness. It is the time to be extra careful about keeping your distance from people with the flu or other viruses. Also, learn how to minimize the effect stress has on your rhythms—and your health—by practicing stress-reducing techniques. (For a list of these techniques see pages 89–90 in chapter 4.)

Unfortunately, the use of chronobiology for preventive medicine is still limited, mainly because we have yet to unlock all the mysterious patterns of our rhythms. Some preliminary studies have demonstrated just how effective rhythms may be as an early warning sign of disease.

In a study of newborn babies, for example, it was found that infants from families with a history of high blood pressure have much larger swings in their daily blood pressure than infants from families without a high blood pressure history.

If such a test became routine, doctors could identify many high-risk babies, without knowing the blood pressure history of the child's family. The high-risk infants could then be raised with full awareness of their potential blood pressure problem and taught to reduce the risk through diet, exercise, and other healthful lifestyle habits.

The rhythmic nature of some diseases has been noted by physicians for centuries. Until recently, however, little was known about how to use these rhythms for prevention and treatment. In the remainder of this chapter, we will look at the major illnesses in which rhythms have been detected, and at how you can use your knowlege of these rhythms to live a longer and healthier life.

ASTHMA

Nearly one in ten of us suffers from asthma, a malady that is usually, though not always, caused by severe allergic reactions to a particular substance—anything from plant pollen to animal hair to a household cleaner. Asthma strikes when the bronchial tubes and membranes of the upper respiratory system swell in reaction to the irritating substance. This causes breathing difficulties and, in severe attacks, a terrifying feeling of suffocation. Medication can usually open up the constricted air passages, but for those suffering from the most severe forms of asthma, an attack can be fatal. Asthma **kills several thousands of people each year.**

Asthma is a notoriously rhythmic disease, with attacks usually coming late at night or early in the morning, when a person is sleeping. This rhythmic pattern was noted and written about by Hippocrates and Aretaeus, two great physicians of ancient Greece.

More recently, researchers have used this knowledge to develop a much clearer picture of the twenty-four-hour cycle of the illness. Here's what they found: The threat of an attack is lowest around three in the afternoon. It remains low well into the evening. Then, around 11:00 P.M., a time when many of us are in the early stages of sleep, the possibility of an attack begins to rise sharply. Early in the morning—between 6:00 A.M. and 7:00 A.M.—the threat peaks. As the morning continues, it begins to decline, eventually reaching its low point again in the middle of the afternoon.

It is this circadian cycle of increasing and decreasing danger that explains why it is not uncommon for someone with asthma to wake in the middle of the night gasping for air but show almost no symptoms during the day.

Doctors have long attributed the nightly nature of asthma attacks to allergic reactions triggered by dust, feathers, or other antigens in people's bedding. It is also widely assumed that the horizontal position of our bodies during sleep causes fluids to pool in our lungs, making an asthma attack more likely.

While these factors may contribute to the onset of an asthma attack, chronobiologists have found stronger—and more rhythmic—explanations for the timing of asthma attacks.

One of the most important is the daily change in the size of the bronchial tubes through which we breathe. In all of us, the bronchial tubes open and constrict on a daily cycle that is remarkably in tune with the cycle of asthma attacks. At about 7:00 A.M., when asthma attacks are most likely, our bronchial tubes are most constricted and breathing is most difficult. At about 3:00 P.M., the low point in the likelihood of an asthma attack, bronchial tubes are at their most relaxed and open.

While this rhythm occurs in all of us, in asthmatics it is more pronounced, and the daily swing between open and constricted is more dramatic. In a nonasthmatic, the difference in the size of the bronchial opening varies only about 5 percent. In people with asthma, the difference is about 20 percent, whether they are suffering from symptoms or not. Indeed, this rhythm is so pronounced in asthmatics that some scientists view it as a way to diagnose the disease when no other symptoms are present.

Chronobiologists have also found that our sensitivity to substances to which we are allergic changes during a twenty-four-hour period in a cycle that coincides with, and may contribute to, the timing of asthma attacks. At about 11:00 P.M. our susceptibility to antigens is at a peak, which means even a small amount of an irritating substance can cause a strong allergic reaction. That susceptibility remains high throughout the night, but declines during the day. By midafternoon, you may not react at all to even large amounts of a substance to which you are allergic.

To complicate matters for asthmatics, researchers have found that some hypersensitive people exposed briefly to an antigen during the day, when their resistance is high, may have a delayed reaction that occurs late at night, when their resistance is low. The person may search frantically around the house, trying to find what it is that is triggering the reaction, not realizing that it was an antigen he or she was exposed to while at work hours earlier.

Like so many other rhythms in our bodies, the cycles of asthma are not simply circadian. Chronobiologists have found that, in women, the frequency and severity of asthma attacks is tied in part to the menstrual cycle. Women asthma sufferers often complain that their asthma becomes worse just prior to and during menses. The reason for this is not clear, but there is some evidence that it is due to monthly changes of hormone levels.

Finally, there is the dramatic yearly cycle in the frequency of asthma attacks, with late summer and autumn being a more dangerous time for asthmatics. Indeed, one study conducted in the Netherlands during an eleven-year period showed an 80 percent increase in the level of asthma attacks between the low point in March and the peak in August. The number of attacks remained high through November, then declined as winter set in.

No doubt, much of this seasonal rhythm is due to the onset of cold winter weather and a drop in the amount of pollen, dust, and other antigens in the air. But many chronobiologists believe that the seasonal cycles of a host of hormones and other chemicals within our bodies also play a part in the annual rise and fall of asthma symptoms.

What you can do. Obviously, you want to try to avoid the antigens that trigger your asthma during the evening hours, when you are most susceptible to an attack. If you are taking drugs for your asthma, remember that they are generally most efficient when taken in the late afternoon or early evening. The drug acetylcholine, for instance, is more effective in making breathing easier when given around 3:00 P.M. The same is true of corticosteroids, hormones often given to treat asthma.

HAY FEVER, HIVES, AND OTHER ALLERGIES

While most people are fortunate enough not to suffer from an allergic disease as serious as asthma, it does seem that almost everybody is allergic to something. There is the rash caused by the neighbor's cat, the hives caused by eating peanuts, or the sneezing fit brought on by household dust.

Think for a moment not only about all the times you have reacted to the particular antigen you are allergic to—say, the dust in your house—but also about all the times you *haven't* reacted to it. You can sit for hours in your living room so absorbed in a book that you don't even notice your surroundings. Suddenly, for no apparent reason, your nose begins to itch and you sneeze. Then you sneeze again, and your eyes begin to water. Pretty soon, things get so bad you can't continue reading.

What happened? What changed to cause your sudden allergic reaction?

You did.

As with asthma, our susceptibility to the things we are allergic to changes dramatically throughout the day. So dramatically, in fact, that one study showed our skin is one hundred times more sensitive to irritants at midnight than it is at two o'clock in the afternoon.

The peak hours of itching, sneezing, and breaking out in rashes are usually in the evening, between 7:00 P.M. and 11:00 P.M.— once again, the time of day when cortisol levels are at their lowest in the body.

What you can do. While it is difficult, if not impossible, to avoid many of the substances you are allergic to, you might be able to reduce your reaction and discomfort by staying away from them in the evening.

If you are allergic to grass and weeds, mow your lawn at 2:00 P.M., when your body's natural defenses are high, instead of in the evening or early morning. If you are allergic to dust or household cleaners, try to schedule big, weekly housecleaning projects for the middle of the afternoon.

COLDS, FLU, AND OTHER COMMON INFECTIOUS ILLNESSES

Every autumn the warning reports appear in the press: Flu season is approaching. Whether it is a new strain or an old one, one from Taiwan or from a country closer to home, the timing of the virus always seems to be the same. It arrives in the autumn as surely as the first chilly winds.

The observance of this annual rite of fall is as old as medicine itself. As the great physician Hippocrates noted some 400 years before the birth of Christ, we are more likely to catch and die from an upper respiratory infection—cold, flu, or pneumonia—in winter than at any other time of the year.

Why this should be is not clear. Chronobiologists are not yet sure whether it is our rhythms, the disease's rhythms, or both that make us more susceptible to a certain infectious illness at a particular time of year.

Many other common infectious diseases also have seasonal rhythms. As parents know, childhood diseases such as chicken pox, mumps, and measles strike more often during the first six months of the calendar year (January to June) than during the last six months (July to December).

What you can do. Infectious diseases are difficult to avoid, mainly because they are usually contagious for several days *before* any real symptoms appear. Still, the illnesses often become even more contagious once the symptoms appear; you should avoid, if possible, people who are exhibiting symptoms. During the fall and winter months, when cold and flu viruses predominate, keep your hands away from your face, nose, and eyes, and wash your hands frequently. Viruses can stay active for several hours on cups, table tops, door handles, and the like. If your hand should touch these objects and then touch your nose or eyes, you can easily become infected. Also, be sure to keep the rooms in your home well ventilated during the flu seasons; fresh air helps disperse the microscopic droplets of virus-laden mucus from a sneeze or cough.

HEADACHES

No pain is more common in modern society than the headache. Indeed, only about one person in ten manages to get through life without experiencing a headache. For the other 90 percent of us, the throbbing pain in the head happens enough to be annoyingly familiar.

Headaches range from those that are extremely painful and debilitating, such as the infamous migraine headaches, to the more common and less severe tension headaches. While all headaches indicate a biological imbalance in the body, less than 5 percent are warning signs of a truly dangerous medical condition such as a brain tumor or meningitis.

More often, headaches are triggered by something more innocuous, such as an allergic reaction to a hot dog or muscle stiffness from sitting in a poorly designed chair. The body tenses its muscles in reaction to the mild toxins of the hot dog or to counteract the strain on muscles from the chair. As the muscles constrict, blood vessels in the head and neck expand, causing irritation to the surrounding tissue. The result is pain; it is known as a headache.

Most headaches are predictable only in their unpredictability. Some, however, do have daily, weekly, or monthly patterns, although the cause of these patterns is sometimes related more to our lifestyle than to internal rhythms.

Tension Headaches

Tension headaches—the most common headaches of the modern world—are caused, not surprisingly, by stress. Because of their connection with stress, they can occur regularly or just occasionally. It really depends on the individual and how stressful his or her life is.

They do, however, seem to strike women three times as often as men, although the reason remains unclear. And they hit city slickers more often than country folk—probably because of the increased stress of city life.

Several studies of the circadian rhythms of tension headaches indicate that they are most likely to strike between 8:00 A.M. and noon. Probably, this is because tension builds in the morning as we confront the problems and stress of the coming day. Tension headaches typically last for a few hours and then fade away.

Researchers have also found that tension headaches occur less frequently on Monday than on any other day of the week. This may well be because Monday falls after the stress-reducing weekend. The number of headaches seems to be spread rather evenly over the other six days of the week.

What you can do. The best thing you can do for a tension headache is to learn how to relax. As soon as the headache begins, lie down, if possible, so you can take the strain off your neck, since this is the location of most of the muscles that tense up and cause a tension headache. Close your eyes and breathe deeply to return oxygen to your strained muscles. Better yet, ask a sympathetic friend or loved one to gently massage your neck and shoulders.

Removing yourself from the source of the stress is also important. If you have just had an argument with a spouse or colleague and feel a headache coming on, go for a walk. Getting outside and breathing fresh air can help get rid of the tension causing the headache.

Taking a couple of aspirin or acetaminophen will reduce headache pain for a few hours, but it won't make the headache go away. If the stress that created the headache is still there when the pain-killing effect of the aspirin fades, the headache will return.

Migraine Pain

Migraine pain begins when the blood vessels carrying blood to the brain suddenly and drastically begin to constrict. Sometimes it brings on a feeling of disorientation, along with hallucinations, numbness, and, finally, pain—excruciating pain.

The migraine headache has cursed humanity for centuries. References to this type of headache can be found in the Sumerian writings of 6,000 years ago. A few centuries later, in 1400 B.C., the Egyptians wrote about developing various medicines to cure it.

The pain of a migraine headache results from a sequence of events that happens in all people, but is exaggerated in migraine victims. The headache typically begins when small blood vessels in the center of the brain constrict, perhaps because of changes in hormone balances brought on by stress.

All of us experience similar blood vessel constriction under stress, but in migraine sufferers, the effect is more drastic. The brain, noting the sudden reduction in life-supporting blood supplies, reacts by opening up, or dilating, larger arteries just under the scalp. In a migraine sufferer, this dilation is highly exaggerated, causing the walls of arteries to stretch. Within the arteries are nerve endings, and as the artery walls stretch, the nerve endings send out excruciating messages of pain. The pulsing, pounding feeling that comes with most migraines is the result of blood being pumped in beats through the arteries.

Migraine headache attacks are famous for their unpredictability, but chronobiologists have noted measurable daily, weekly, and (in women) monthly rhythms.

Different studies have pinpointed different hours as the most likely time for a migraine attack. Some put the time between 4:00 A.M. and 8:00 A.M.; others say the peak period for a migraine is later in the morning, between awakening and noon. All seem to agree, however, that the morning hours are the worst for migraine sufferers. Migraine sufferers frequently rise from their slumber to the pounding onset of a miserable and debilitating headache.

Migraine headaches also display a weekly rhythm, although the reason for this is not well understood. In a study conducted in the early 1970s, the migraine patterns of 2,933 women were analyzed. Researchers discovered that the women had the highest number of migraines on Saturday and Sunday, and the fewest on Monday. A later study cited Thursday and Friday, not the weekend, as the peak days for migraines, but the low point remained Monday.

Originally, this weekly headache cycle was attributed to weather conditions, but closer study showed that this explanation was unlikely. Some researchers now believe the weekly rhythm may be connected to the increasing and decreasing emotional stress of the work week and the weekend, but that has yet to be proven.

Some chronobiologists, on the other hand, believe that the weekly migraine cycle may not have an outside cause at all, but instead may be yet another example of one of the natural seven-day cycles that occur in so many biological functions. (See chapter 1.) No direct evidence has been found, however, to support this theory.

The monthly rhythm of migraine headaches in women has long been known, and it is tied closely to menstrual cycles. Several studies have shown that more than a third of the migraine attacks that afflict women occur during the four days before menses. About another third occur during menses. Surprisingly, women on the Pill are also more likely to have a migraine attack just before and during their periods.

What you can do. After centuries of searching, physicians discovered in the 1930s that a drug called *ergotamine*, given at the right time and in the right amount, can stop a migraine attack. It works by constricting the blood vessels that cause the pain of migraine.

You should be aware that ergotamine often causes severe nausea; but, because migraines also cause nausea, this side effect is considered acceptable by most doctors and migraine sufferers. Ergotamine can be administered in a variety of ways, from injection by needle to nasal spray inhalers. The drug even comes in a suppository, which enables it to enter the bloodstream without going through the stomach.

While you can treat the pain of a migraine, a cure has not been found. Remember, a migraine headache is a physical illness caused by a nervous system that cannot properly control the expansion and contraction of blood vessels. It often runs in families (about 70 percent of those who suffer from migraines inherited the illness), and there seems to be little you can do to cure yourself of the illness.

You can, however, lessen the chance of an attack by identifying and avoiding some of the objects or situations that can trigger a migraine—especially in the morning or, if you are a woman, just before or during your menstrual period. Common triggers include food allergies (to milk, chocolate, eggs, wheat, coffee, or alcohol, for example), cigarette smoke, perfumes, dust, bright light, loud noise, and emotional stress. After each attack, analyze what you were doing or eating right before the headache began to see if you can find a pattern and pinpoint exactly what set it off.

Pain in Clusters

The most rhythmic of all headaches, as well as the most painful, is the cluster headache. It is actually a series, or, as its name denotes, a cluster of headaches that strike with clocklike precision at certain hours of the day during certain times of the year—all of which vary from person to person. It is also the only type of headache that strikes men more frequently than women.

That biological rhythms control the occurrence of cluster headaches is in little doubt. Which rhythms are involved and how they work, however, has not been determined.

Cluster headaches, like migraines, involve the constriction and dilation of blood vessels, but for some reason they are more painful than migraines. Those who suffer from them can often predict fairly accurately when they will hit, when they will stop, and when they will begin again.

For some the pattern may be an hour-long headache that recurs at the same time of year, three times a day for several weeks, then goes away until the same period a year later. Others may have a two-hour headache once a week for six weeks, then no headaches for another year. Occasionally, the headaches vanish altogether; they may stop for several years before returning.

Studies by chronobiologists have found that cluster headaches strike most often while we sleep. One study showed that 60 percent of cluster headaches begin between 11:00 P.M. and 6:00 A.M. Many patients in the studies said the headaches seemed to start about one to two hours after they went to sleep—or just around the onset of the first REM, or dream, period of their nightly sleep cycle.

Cluster headache victims apparently don't inherit the illness, a difference from many migraine sufferers; but, like migraine victims, they do have the same trouble dealing with stress. Cluster victims also tend to drink more coffee and alcohol, and to smoke more than other people.

Scientists have focused their search for the cause of cluster headaches on an imbalance in the daily ebb and flow of hormones in the brain that appear to trigger the constriction of blood vessels that lead to these headaches. As yet, however, they have been unable to link hormonal cycles directly to the headaches.

What you can do. To ease the pain of cluster headaches, you should stop drinking alcohol, cut down on smoking, and in general take better care of yourself during a cluster period. Although there is little evidence that discontinuing smoking or alcohol consumption will have a long-term impact on the headaches, many victims are painfully aware that a cigarette or a glass of wine can trigger a sudden headache.

For extremely severe cases, surgery is sometimes performed to cut nerves or freeze tissues that contribute to the headaches. These procedures are risky and should be attempted only as a last resort.

Finally, there are strong prescription drugs, the most common being *methysergide*. Unlike ergotamine, which is taken just as a migraine headache strikes, methysergide is taken several times a day during the cluster period to prevent or lessen the pain of the headaches.

When to Visit the Dentist

Nobody really *enjoys* going to the dentist. The shrill sound and grinding vibration of the high-speed dental drill as it works on a cavity-ridden tooth is enough to make even the strongest among us shudder.

But, take heart. Your daily rhythms can help you.

Teeth have a pain threshold that rises and falls each day. You are most sensitive to a painful tooth in the early morning, between 3:00 A.M. and 8:00 A.M. So, if your tooth hurts the evening before you go to bed, you're probably in for a long night, because the pain is only going to get worse as dawn approaches.

If you should wake up with a toothache, think twice before rushing immediately to the dentist. The bad tooth is also more sensitive to the pain of drilling and grinding at that hour. Midafternoon is the best time for scheduling dental work, because your threshold for tooth pain rises as the day progresses, reaching a peak around 3:00 P.M. At that hour, your pain threshold is *50 percent higher* than it is early in the morning.

The effectiveness of novocaine and other numbing drugs used in dentistry also varies dramatically with the time of day. Anesthesia given in the afternoon lasts much longer, sometimes several hours longer, than it would if it were given early in the morning.

So if you can put off seeing your dentist until midafternoon, you may find the trip a little less painful.

Changes in behavior, such as reducing stress, or even biofeedback, do not seem to have much effect on changing the rhythm of cluster headaches.

ULCERS

Out-of-sync biological rhythms, perhaps induced or made worse by stress, appear to play an important role in the development of ulcers.

Scientists now know that the gastric acids in our bodies—the juices in the stomach and small intestine that help break down and digest the food we eat—have a circadian rhythm. In general, our bodies put out little gastric acid during the night, when we are sleeping. During the day, the amount of acid gradually increases, reaching a peak around midday.

In many people who suffer from ulcers—a stomach ulcer or a duodenal ulcer (an ulcer of the small intestine)—the rhythm of acid production malfunctions. First, its daily peak lasts longer than normal. This means that more acid remains in the stomach and intestines. At night, when gastric acid production should be low, it stays high.

Changes in the rhythm of acid production can leave the stomach and intestines unprotected from the corrosive effect of the acid. Why? Because other secretions that protect these internal organs from the acid do not switch *their* rhythms. The mucus that coats and protects the small intestines, for example, has a rhythm that runs just ahead of that of the gastric acid. If the production of gastric acid suddenly shifts into high gear at night, the production of mucus does not necessarily do the same.

The result: Without the protection of the mucus, the acid eats away at the lining of the small intestine, causing an ulcer.

Indeed, one study found that a group of patients with duodenal ulcers had as much as twenty times more gastric acid in their bodies at night than did a healthy control group. Interestingly, it was also discovered that the gastric acid was produced in bursts

While in the Hospital

Going into hospital can often be just as hard on your rhythms—and, thus, on your health—as traveling long distances or keeping an erratic schedule. That's because hospitals have their own set of rigid time cues—mealtimes, medication times, bedtimes—that can throw your rhythms a quick and confusing curve.

On the other hand, some hospital settings, particularly intensive care units, are *rhythm-free*. In other words, these units have almost continuous light, noise, and activity. This absence of strong zeitgebers can also throw your rhythms into confusion and disarray.

If you are going to be in the hospital for a short time and if you want to minimize the effect of the stay on your rhythms, you should try to stick to your usual daily routine as closely as possible while in the hospital. This means eating and sleeping at your usual hours. For maintaining your sleeping hours, you may need to use some gentle assertiveness—to persuade your roommate, for example, to turn off the television set at your bedtime. To maintain your mealtimes (if your medical regimen permits it), you may need your own source of food—a friend or relative, perhaps, who can bring you healthful sandwiches, fruits, and juices that can be stored safely without refrigeration until the appropriate hour. (Considering the poor nutritional and gastronomic quality of much of today's hospital food, you'll probably be better off eating your own food, anyway!)

Of course, if your doctor has you on a special diet, be sure to check with him or her first before planning your own menus.

that coincided with REM sleep. These bursts may be connected to the bursts in hormone activity that often occur with REM sleep.

The daily rhythm is not the only rhythm associated with ulcers. One scientist has discovered that ulcers are more likely to occur in the fall—specifically, September and October—than at any other time of the year. A second, smaller peak occurs in January. Scientists do not know why. Even more mysterious is the fact that most of the ulcers that occur during these peak periods do so when the moon is full! Why is anybody's guess, but perhaps moonlight sets off some kind of change in the gastric acid cycle.

What you can do. A study has shown that if rats are put under a great deal of stress at the time when their gastric acid cycle is at its daily peak, they are more likely to develop ulcers than if stressed at any other time of the day.

The researchers who conducted the study believe its findings bear a warning for humans as well as rats: Daytime stress, especially when it is experienced around midday, when gastric acid is at its peak, can cause more acid than normal to be produced, increasing the chance of its burning a hole in your stomach or small intestine.

One of the researchers suggests that people who suffer from stomach or intestinal problems schedule high-pressure business meetings or other tense events late in the afternoon, after the gastric acid peak has passed. If you are on medication for ulcers, the medicine will probably be most effective if taken in the late morning, just before the gastric acid climbs to its daily peak.

HIGH BLOOD PRESSURE

High blood pressure, or hypertension, batters the interior walls of our blood vessels, forces our hearts to work too hard, and causes fatty plaque to clog our arteries. It does so quietly over many years, often without any overt signs of the damage it is causing. Then, finally, it mounts its strongest, and often fatal, assault on the body—usually in the form of a stroke, a sudden heart attack, or a failed kidney. No wonder it is widely known as the silent killer.

In most modern nations, up to 18 percent of the population may have high blood pressure. In the United States, the total number of people who suffer from the disease is estimated at about sixty million.

What causes high blood pressure? The underlying cause of the illness remains a mystery, but lifestyle seems to play a major contributing role. Studies have shown that if you are overweight, eat too much salt, or lead a sedentary lifestyle, your risk of high blood pressure increases dramatically. However, not everyone with unhealthy lifestyle habits comes down with the disease. Scientists do not know why.

From a purely biological point of view, your blood pressure is regulated by the size of your blood vessels, the daily rhythms of your hormone levels, and nerve cells known as *baroreceptors*. The latter act like thermostats, regulating the pressure of blood as it flows through your body. In people who suffer from high blood pressure, some part of this regulatory system goes awry.

What is a safe blood pressure level? Most scientists agree that a blood pressure reading of 120/80 is normal in a healthy adult and anything above 140/90 is dangerously high.

The two numbers represent the high and low of your blood pressure as your heart beats, then relaxes, then beats again. In the case of the normal blood pressure, the 120 is the pressure measured when your heart is contracting and exerting the greatest pressure on your circulatory system. It is known as the *systolic* pressure. The lower number, known as the *diastolic* pressure, is the pressure measured when your heart is between beats and relaxed.

Unfortunately, most of us pay little attention to our blood pressure and have it checked only when we happen to visit our doctor's office, usually for some unrelated medical problem. Considering the devastating effect high blood pressure has on our health, as well as the frequency with which it occurs, this casual approach to keeping tabs on blood pressure is foolish, at best.

The value of an occasional blood pressure check has been called into question by many chronobiologists. Your blood pressure, like your temperature, *rises and falls* during a twenty-four-hour cycle. Taking a single measurement gives you only a snapshot of where your blood pressure happens to be at that moment in the daily cycle, not an accurate picture of your overall cycle.

For most of us, our blood pressure begins to rise early in the morning, an hour or so before we awaken, and then continues to rise throughout the day, peaking in the late afternoon or early evening. It falls more quickly than it rises, hitting its low point around midnight and staying there until early morning, when the cycle begins again.

If your blood pressure is checked in the early morning, it will be considerably lower than it is in the evening. Indeed, the difference between your low and high readings over the course of a day can be so great that the highest measurement of your relaxed, or diastolic, pressure can actually exceed the lowest reading of your contracted, or systolic, pressure. The "snapshot approach" to taking blood pressure also misses the extent of the daily swing in your cycle.

Your blood pressure may be well below dangerous levels throughout much of the day, but may rise to very high levels for an hour or two in the evening. If this swing goes unnoticed and untreated for many years, it could eventually kill you. Unless you have your blood pressure taken in the evening, during the hours when it is above the safe limits, you have no way of knowing the danger you are in.

The typically haphazard way most of us keep track of our blood pressure also makes it very difficult to notice when the overall range of our daily cycle begins to inch upward on the pressure scale as we get older. Thus, we can easily miss an opportunity to bring our blood pressure under control before it does any damage.

The importance of keeping your blood pressure within safe limits cannot be overstated. Even a person with mild high blood pressure—not too far above the 140/90 threshold—has *double the chance* of dying by age sixty-five than someone with normal blood pressure. As your pressure goes up, so do the odds that your life will be a short one. Someone with just moderate high blood pressure, say 150/100, is *three times more likely* than a healthy person to die by age sixty-five.

What you can do. The first step in protecting yourself against the ravages of high blood pressure is to determine the range of your blood pressure. To get a reasonable picture of the extent to which your blood pressure rises and falls each day, chronobiologists recommend that you measure your pressure every hour or so for two days. (See box on page 161.)

Measuring your own blood pressure isn't difficult, and the equipment you need—a stethoscope, a sphygmomanometer, an inflatable cuff, and a pressure gauge—is available, complete with directions, at most pharmacies. There are also some electronic devices available that, while more expensive, make the task a bit easier. Talk with your physician or pharmacist about which equipment best suits your needs and budget.

If you have a family history of hypertension or close relatives who died from heart attacks, congestive heart failure, or strokes, you might want to take more detailed readings of your blood pressure. To do that, try to do what chronobiologists do when they are measuring their blood pressure: Wear a small, portable monitoring device that automatically takes blood pressure readings at preset intervals throughout the twenty-four-hour cycle. The computerized devices are expensive, so buying one is not practical. But if you live near a large hospital or university, you may be able to borrow one of the devices for several days. Ask your physician if he or she can arrange the loan of such a device.

When analyzing your blood pressure readings, remember that your blood pressure does not have to be above the safe range for very long each day to cause a problem. If you find your pressure regularly straying into the hypertense range—*even for a few minutes a day*—bring it to the attention of your doctor immediately.

Children can also have high blood pressure, especially if they are from a family with a strong history of hypertension. Indeed, doctors have identified hypertension in children as young as twelve months old. High blood pressure is of even greater concern in children than adults because it starts doing its damage at an early age and can cause serious medical problems by the time the children become young adults. In families with a history of high blood pressure, it is important to include the children when monitoring for hypertension.

What should you do if you have high blood pressure?

You can take several simple steps to help bring your blood pressure back down into a safe range:

• *If you are overweight, shed some pounds.* Even being ten or twenty pounds above your proper weight can add stress to your heart. When you are overweight, your heart is forced to pump a greater volume of blood through a larger body.

• *Reduce the amount of salt in your diet.* Most people consume three to five times more salt than they need. Much of that excess salt comes directly from the salt shaker, but hidden salt in processed foods also contributes to our overconsumption of this nutrient. Throw your salt shaker away (enough occurs naturally in foods to meet your daily needs) and season your food instead with herbs, garlic, spices, and pepper. Also, learn how to read food labels; look for the amount of sodium listed.

• *Exercise regularly.* Aerobic exercise—such as running, swimming, cycling, skating, and fast walking—can lower blood pressure. Besides helping to control your weight, aerobic exercise makes the heart and other muscles of the body more efficient, which means they require less blood when under stress.

For an aerobic exercise to work, it must tax your body *continuously* for at least twenty minutes. That's why run-and-stop sports, such as tennis, racquetball, basketball, and even soccer, should not be considered aerobic exercise.

Be sure to consult with your physician before starting an exercise program. Some doctors recommend that people with high blood pressure avoid isometric sports, such as wrestling, waterskiing, and weight lifting, which tighten muscles and actually raise blood pressure.

• *Stop smoking*. The nicotine in cigarettes, by blocking certain nerve impulses, causes the muscles in your small blood vessels to constrict. The result is an increased resistance to blood flow, forcing your heart to work harder and your blood pressure to rise.

Tracking Your Blood Pressure

To track the daily rhythms of your blood pressure, you will need to obtain a portable monitoring device, either through your physician or local pharmacist. Read the directions for using the device carefully. Better yet, have your physician or other trained specialist instruct you in its use.

To get a clear picture of the daily cycle of your blood pressure, you should take a reading every hour for at least two days. Then record those readings on the charts on the following pages. Be sure to include both your systolic pressure (the pressure of your blood when your heart pumps) and your diastolic pressure (how far the pressure drops between heartbeats). For easier reference, use a different colored pencil or pen to record each type of pressure.

At the end of each day, "connect the dots" for each type of pressure and note if they both fall within the safe ranges indicated on the chart. If any of your readings do not fall within these ranges, then you should contact your physician for a more thorough blood pressure evaluation.

HEART ATTACKS

If you are going to have a heart attack, chances are it will strike in the morning. The number of heart attacks that occur between 8:00 A.M. and 10:00 A.M. is about double the number of attacks that occur during the evening or late night hours.

While there are many things that increase your vulnerability to

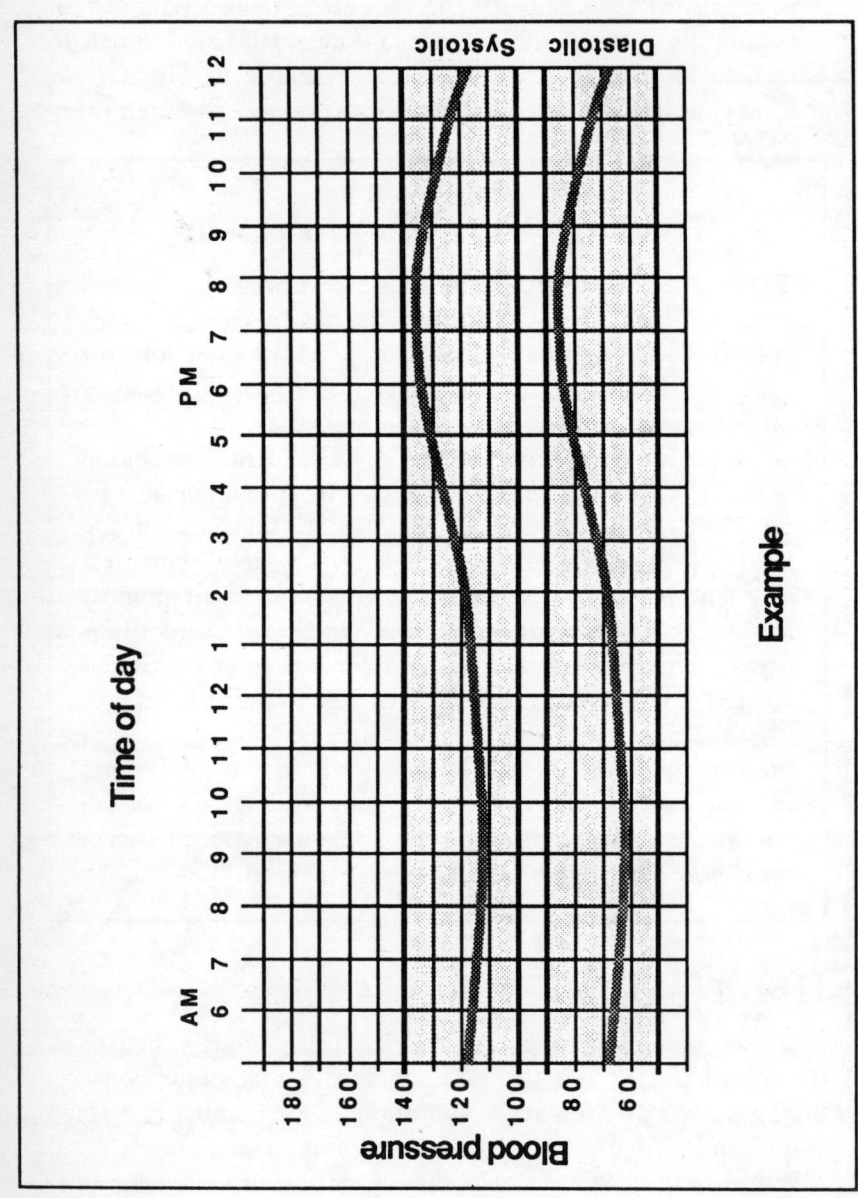

Example

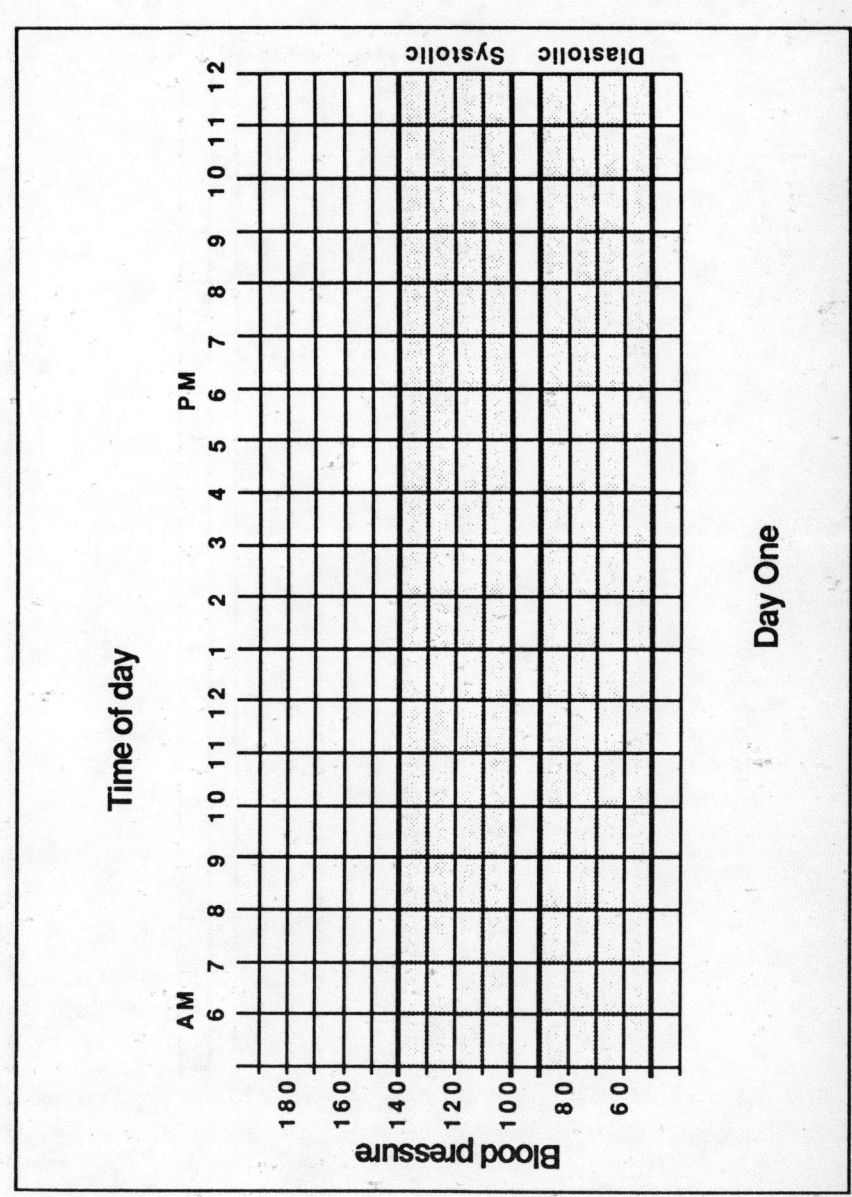

Time of day

Blood pressure

Day One

Diastolic   Systolic

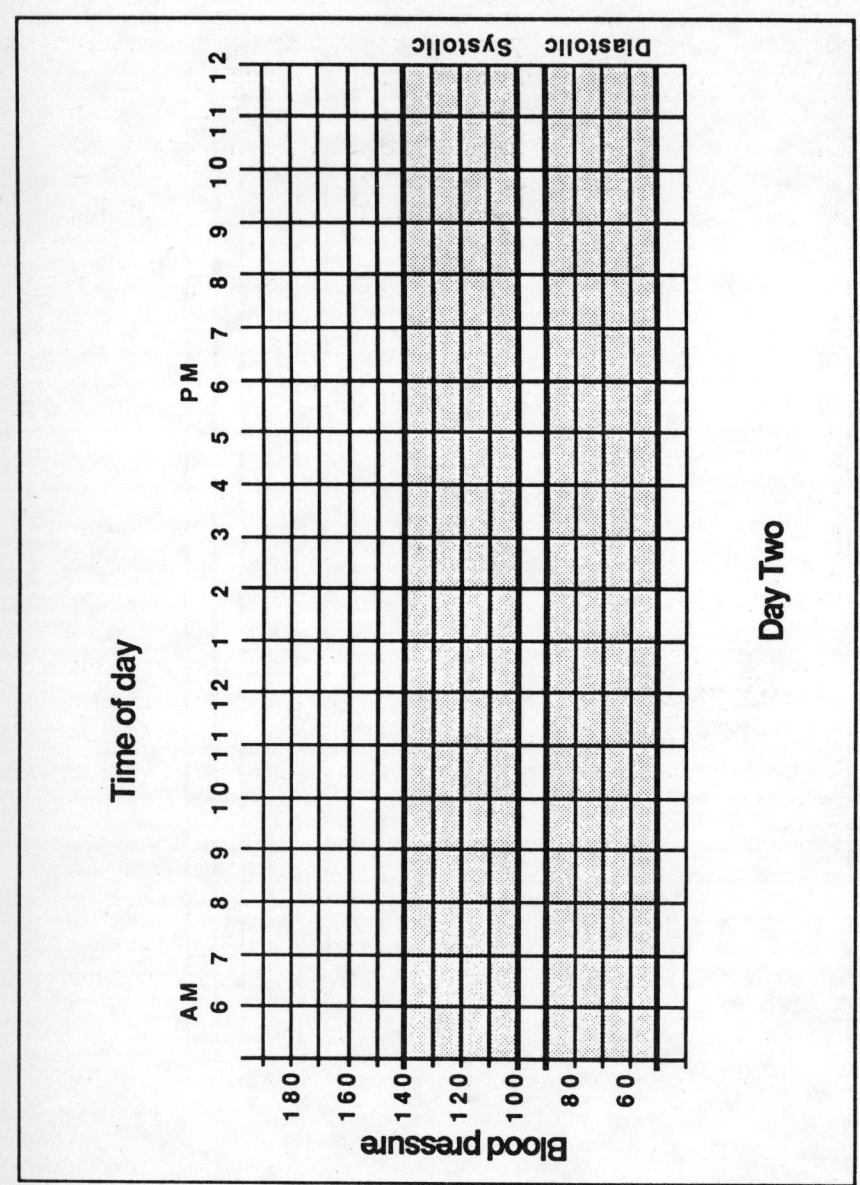

Day Two

a heart attack—smoking, diet, stress, and excess weight—chrono-biologists believe the daily rhythms of your heart also play a strong role in the heart's susceptibility to failure.

The daily fluctuation in blood pressure is just one of these rhythms. Scientists have also found circadian rhythms in the heart's stroke rate, in the amount of blood it pumps, in the ease with which blood flows through blood vessels, and in other more subtle functions of the heart.

Even the rate at which your heart beats follows a clear daily rhythm. The heart generally beats faster during the day than at night, closely following the daily ups and downs of body temperature. In fact, before the oral thermometer was invented, doctors used pulse rate as a rough guide for tracking a fever. A heart rate increase of ten to fifteen beats per minute was considered a sign that the patient's body temperature had risen one degree.

Although heart rate is affected by temperature, it is not dependent on it. After a heart transplant, the new heart and the tissue that remains in the body from the old heart run on different circadian cycles for some time. In other words, they beat at different rhythms! Often the hearts are more than two hours out of sync with each other. The fact that the tissue from the old heart holds so strongly to its old rhythm indicates to scientists that the pacemaker determining heart rate is not based at a central location in the body but in the individual cells of the heart itself.

All these various rhythms work together to change the heart's efficiency during the day and, thus, its susceptibility to failure. Treadmill studies have shown that the heart seems most able to tolerate heavy stress around 5:00 P.M., and least able to handle it around 9:00 A.M., the time that coincides with the peak in heart attacks.

So just when we are becoming active in the morning and bracing ourselves to face the physical and emotional stresses of the day, our hearts are at a circadian low in their efficiency and are at their most vulnerable.

What you can do. If you have a history of heart attacks in your family or have an increased risk because of smoking or poor diet, it is important to remember you are most vulnerable to problems in the morning. Physical exercise, such as running, should be done in the afternoon or evening. Highly stressful business meetings should be scheduled for the afternoon, not the morning, whenever possible. Try to ease yourself into the day, letting your body slowly warm up.

Of course, reshuffling your daily activities will not, by itself, decrease your risk of a heart attack. Although heart attacks occur most often in the morning, they can occur at other times, too. A better bet is to talk to your doctor about steps you can take, such as stopping smoking, that will make your heart less vulnerable to the body's natural early morning circadian sag.

A PAINFUL HEART

Many people suffer from a disease known as angina pectoris, or chest pain. When this disease strikes, the small arteries that supply blood to the heart suddenly contract, resulting in severe pain. Angina is often caused by physical stress. The pain can usually be stopped by relaxing the body or by taking nitroglycerin, which reopens the arteries.

Chronobiologists have found that the likelihood of an angina attack is tied to the changing tone of the arteries, which has a daily rhythm. In the morning, the arteries of people suffering from angina are constricted, and exercise can easily trigger an attack. In the afternoon, the arteries allow more blood to flow through and thus can tolerate more exercise.

In one study, thirteen patients with angina pectoris exercised on a treadmill between 5:00 A.M. and 8:00 A.M. All of them suffered episodes of chest pain. When the same group of patients was put through a more stressful exercise program, but this time between 3:00 P.M. and 4:00 P.M., only two suffered an angina attack.

What you can do. Clearly, if you suffer from angina, you should exercise in the afternoon or evening, but not in the morning. If you take nitroglycerine, a common medicine for a host of heart ailments, you should be aware that the medication is more effective in the morning, when blood vessels are most constricted. Whether the same holds true in those suffering from other forms of cardiac disease is not known, but it is an issue worth discussing with your doctor.

CANCER

Almost all of the cells in our body divide to a rhythmic daily beat. Cells in contact with the outside world, such as those in our skin and eyes, tend to divide late at night. The cells of some other organs, such as the adrenal gland, tend to divide during the day, when we are most active. In fact, each organ of our body has its own daily rhythms of cell division, and scientists are slowly beginning to recognize the tremendous significance this has for cancer therapy.

While most healthy cells divide at their regular, appointed time each day, cancer cells do not. They divide erratically, with no sense of time. If they do develop a rhythm, it is an abnormal one—such as dividing twice a day instead of once.

In any event, they are cells out of sync with the healthy tissue that surrounds them. Chronobiologists are now trying to take advantage of that fact to offer better treatment to cancer patients. Experts have found that cancer cells, because their rhythms differ from the healthy cells around them, are vulnerable to specific, timed attacks of radiation or anticancer drugs.

Several examples of the importance of drug timing in cancer treatment have been given earlier in this chapter. (See sections on using rhythms to detect and treat diseases.) Here is one more: In 1985, a group of chronobiologists at the University of Minnesota Hospital began a study of thirty-one women suffering from advanced ovarian cancer. The women were divided into two groups and treated with similar doses of two of the most common, yet toxic, anticancer drugs, adriamycin and Cisplatin. Treatment differed only in the timing of the two drugs.

The first set of women, Group A, received adriamycin in the morning and Cisplatin in the evening. The second set, Group B, received the drugs in the opposite order—Cisplatin in the morning and adriamycin in the evening.

The reactions of the two groups to the drugs were remarkably different. Group A had fewer complications, fewer dose reductions because of bad reactions to the drugs, and fewer treatment delays because of physical problems. In other words, a *morning dose* of Cisplatin appears to be *more toxic* to healthy body cells than an evening dose. The reverse is true for adriamycin: It appears much *less* toxic when taken in the morning.

The women in Group B, because of the complications and toxic reactions they had to the drugs, received less intensive therapy for their cancers than did the women in Group A. And that can sometimes make the life-or-death difference in fighting cancer.

The study did not look at the curative effect the timing of the drugs had on the cancerous tissue, but another study did, and it also came up with dramatic results. In this second study, twelve women, all suffering from advanced ovarian cancer, were put on a carefully timed treatment schedule based on their individual circadian rhythms.

After nine months of timed chemotherapy, the cancer in *ten* of the women *disappeared*. Later surgery on four of the women confirmed that the cancer was indeed gone. The remaining two women were in partial remission for several months. That translates to an 83 percent success rate in treating this often deadly form of cancer.

Compare that with another group of nineteen women who also suffered from the same type of ovarian cancer. Although they were given the same doses of the same drugs as the twelve women in the other study, only 20 percent of this second group experienced a complete remission of their cancer. What was different? They received the drugs at random hours, without regard to their circadian rhythms.

Unfortunately, this kind of research has been conducted using only a few of the more than fifty types of anticancer drugs. Still, chronobiologists believe strongly that drug timing holds great hope for the future.

They also believe that one day they may be able to manipulate cell division through diet or other techniques to make sure that healthy cells are least vulnerable to chemotherapy at the time when cancer cells are most vulnerable. That work, however, has a long way to go.

The Seasons of Cancer

A large-scale study conducted in 1984 found that three types of cancer—breast cancer, prostate cancer, and a common type of testicular cancer known as *seminoma*—are most likely to strike at particular times of the year. Scientists believe these seasonal rhythms are related to the annual cycles of sex-related hormones. The hormones may suddenly cause the cancers to grow rapidly—which makes them more likely to be noticed.

Here's what the study found:

• *Breast cancer, one of the leading causes of cancer deaths among women, is most often discovered in the spring.* Doctors diagnose about 30 percent more cases of this disease in May, the peak month, than in December, the lowest month.

• *Prostate cancer, one of the leading causes of cancer deaths among men, is most often diagnosed in late winter and spring.* The least likely month for the cancer to be diagnosed is August, when the number of new cases is 40 percent below average. The diagnosis of prostate cancer rises in March to about 40 percent above average and stays close to that level until May.

• *Seminoma testicular cancer is most frequently diagnosed during the winter months.* In fact, a diagnosis is 85 percent more likely in winter than in summer. Interestingly, another form of the illness, known as *embryonal testicular cancer,* is diagnosed more often in the summer.

What you can do. If you have been diagnosed as having cancer and you face radiation or chemotherapy treatment, you should raise with your doctor the issue of timing your treatment based on the rhythms of your healthy cells.

Unfortunately, not all physicians are familiar with timed treatment. Dr. William Hrushesky of the University of Minnesota Hospital, one of the physicians leading the chronobiology cause, recommends that patients educate themselves about timed treatment before talking it over with their doctors. The place to begin is at your local medical library. (Most universities have medical libraries.) Ask the librarian to help you find the most recent cancer studies relating to chronobiology.

People who are at high risk for breast cancer, prostate cancer, or testicular cancer should also alert themselves to the seasons when these illnesses are most likely to occur. (See box on page 169.) One of the most important aspects of successfully treating cancer is discovering it early.

Women, therefore, should be extra alert in the spring for unusual breast lumps. And men should be extra alert during the winter and spring months to painful testicles or problems with urinating—early signs of testicular and prostate cancers.

EPILEPSY

Epilepsy is a disease marked by rhythms. For more than 150 years, scientists have noted that epileptics tend to develop individual patterns, or cycles, to the timing of their seizures.

While these cycles aren't precise enough to enable an epileptic to predict seizures, they are regular enough to allow researchers to develop categories for epilepsy sufferers based solely on the timing of their attacks. These categories are:

• *Diurnal* epileptics—about 40 percent of all epilepsy victims—experience their epileptic episodes during the daytime hours.

• *Nocturnal* epileptics—about 25 percent of epilepsy victims—have most of their epileptic attacks at night.

• *Diffuse* epileptics exhibit no day or night pattern to their attacks; they experience them at any hour of the day.

• *Awakening* epileptics, which make up a very small percentage of all epilepsy victims, tend to suffer their seizures soon after they wake up in the morning.

The timing of epileptic seizures is thought by some researchers to be closely related to the sleep-wake cycle because the timing of the seizures can be shifted forward or backward by making a similar shift in the timing of the sleep cycle.

For women who have epilepsy, the incidence of seizures also follows a monthly cycle tied closely to the menstrual cycle. The likelihood of a seizure increases dramatically a few days before menstruation, peaks on the first day of menses, then stays high until menses ends.

As with all the rhythms of epilepsy, scientists are not sure what causes the increase in seizures during menstruation. Some believe it is related to the ebb and flow of the reproductive hormone progesterone, but there are no conclusive answers.

What you can do. Knowing when a seizure is likely to strike can be immensely helpful to an epileptic, who can then use this information to schedule important events. Unfortunately, seizures are infamous for striking when they are least expected, so knowing when your seizures are most likely is no guarantee that they will not occur at another time of the day. Still, it puts the odds a bit in your favor.

Understanding the underlying rhythms that lead to your seizures can also help you and your physician time your anticonvulsive drugs so they are taken at the time of day when they will be most effective. One study of a group of epileptics, for example, found that brain-wave abnormalities occur at the time of day when an individual is most likely to suffer from an attack. Interestingly, those abnormalities occur at the same time, day in and day out, whether or not the person suffers an attack. This daily rhythm proved to be a deep-rooted constant for as long as a decade in several of the patients studied.

SEXUALLY TRANSMITTED DISEASES

Syphilis and gonorrhea, two sexually transmitted diseases that are epidemic among young adults, also have seasonal cycles. Reports of gonorrhea peak in late summer; of syphilis, in late autumn. These rhythms are particularly interesting because they fit neatly with the annual rhythm of human sexual activity. Here's why:

Humans are most sexually active in late summer and early autumn.

Gonorrhea symptoms usually appear anywhere from two days to three weeks after exposure to the disease.

Syphilis symptoms generally take longer to develop, anywhere from nine to ninety days after exposure.

It makes sense, therefore, that reports of gonorrhea are greatest around the time of peak sexual activity (late summer) and that reports of syphilis peak a few weeks later.

Genital herpes, a sexually transmitted virus that can be controlled but not cured, does not seem to have an annual season. That may be due as much to record keeping as it is to the diagnosis of the disease, however. Unlike gonorrhea or syphilis, doctors are not required to report cases of herpes to state health departments, so it is impossible for chronobiologists to get an accurate picture of the disease.

In some women infected with genital herpes, the outbreak of the sores and blisters associated with the disease seems to be tied to their monthly menstrual cycle. This monthly recurrence of the blisters seems to be related to both the physical and emotional stress menstruation causes in some women. The outbreak often occurs during or just after menstruation.

AIDS, or acquired immune deficiency syndrome, the newest and most deadly of the sexually transmitted diseases, has yet to show any indication that it has seasonal cycles.

What you can do. Given the life-and-death seriousness of AIDS, the incurable nature of herpes, and the health risks—including infertility—of other sexually transmitted diseases, you should always practice "safe sex."

The best defense against sexually transmitted diseases, of course, is a long-term, monogamous relationship.

If you are meeting and seeing new people, however, there are precautions you should take to protect yourself. First, get to know the other person before becoming intimate. Whether you are male or female, casual, indiscriminate sex with a variety of partners is the quickest way to expose yourself to a host of sexual diseases.

If you do become intimate with someone you do not know well, protect yourself. If you are a man, use a condom. If you are a woman, insist that the man use a condom. While condoms are not a guaranteed defense against sexually transmitted diseases, they are the best protection available.

Tips for Keeping a Healthy Beat

• Chart your daily temperature rhythm so you can use it as a control to determine whether or not you are coming down with an illness.

• Periodically chart your daily blood pressure rhythm to make sure your pressure is within safe limits. It will also help you keep track of any upward—and potentially dangerous—climb in your blood pressure as you age.

• If you suffer from a particular illness, learn the rhythms of that illness so you can take medications or other treatments when they will be most effective.

• Try to schedule dental appointments in midafternoon, when your threshold for tooth pain is at its highest.

Seven

~

You Are When You Eat

Tell me what you eat, and I will tell you what you are.
—ANTHELME BRILLAT-SAVARIN

We are not only what we eat, but when we eat.
—FRANZ HALBERG

Early in October, on her thirty-fifth birthday, Judy's husband surprised her with two plane tickets to Jamaica. They would be going around Christmas, he told her, for a full week of sun, sailing, and sandy moonlight walks—just the two of them. The kids would stay with grandma.

Judy was excited about the prospect of a second honeymoon—and was pleased that the trip was more than two months away. She would have plenty of time to lose those seven pounds she had been meaning to shed. A couple of years earlier, Judy had successfully lost ten pounds after the birth of her second child. That diet, she remembered, had been fairly easy. For three months, April through June, she had simply cut down on the amount of food she ate and had increased her exercise slightly. The pounds had disappeared painlessly.

Judy decided to try the same approach. This time, however, the diet didn't go as smoothly. For one thing, Judy found it much harder to cut back on calories. In fact, if anything, she wanted to eat more! During those weeks when she was able to control her urge to eat, she didn't seem to lose as much weight as she had on the earlier diet.

By the time she boarded the plane with her husband for Jamaica in December, Judy had lost some weight—but only a disappointing three pounds.

Although Judy didn't realize it, her trouble with dieting had more to do with the season than a lack of determination. Judy was trying to take off weight in the autumn, and most of us find it easier to take off weight in the spring than in the autumn. The reason lies with our biological rhythms, for we are metabolically different in the autumn than we are in the spring.

We are also much different in the morning—metabolically speaking—than we are in the evening. That, too, has important consequences for people who are trying to lose or gain weight.

Your Eating Habits: A Quiz

- Do you often skip breakfast?
- Do you eat fatty foods for breakfast—things like doughnuts, pastries, buttered toast, and fried eggs?
- Morning people only: Do you drink two or more cups of coffee or tea in the morning?
- Do you eat bigger meals as the day progresses?
- Do you often skip lunch?
- Do you drink coffee or tea throughout the day?
- Do you drink caffeinated beverages (including cola-type soft drinks) after 6:00 P.M.?
- Do you have a cup of cocoa or milk before going to bed at night?
- Do you cut down on physical activity once summer is over?

If you answered yes to any of these questions, then you are not using food to your best "rhythmic" advantage.

This chapter will look at the relationship between food and our biological rhythms. It will explain how you can use your knowledge of rhythms to help you lose weight. It will also show you how you can improve your daily mood and performance rhythms by carefully selecting and timing the foods you eat.

WHY YOU PUT ON WEIGHT IN THE AUTUMN

Although you are probably unaware of it, your eating habits have a strong seasonal pattern. In simplest terms: You eat more in the autumn and winter than in the spring and summer.

Your body also is more adept at turning calories into body fat during autumn and winter, so it should come as no surprise that you are more likely to gain weight during those two seasons than at other times of the year.

Why this seasonal upswing in our desire for food? It has to do with the preservation of our species. Our ancestors needed an extra layer of fat on their bodies to help ensure their survival through the cold and food-scarce winter. Their biological rhythms "conspired" to make them eat more and store it more efficiently as soon as the days grew shorter every autumn.

Unfortunately, biology has not yet caught up with central heating and all-night supermarkets. Each autumn we put on a few more pounds—weight that many of us could do without and that most of us find hard to lose when spring arrives.

You can fight back, however. Start by being alert to an increase in your appetite during the autumn months. Think *ahead of time* of eating less at that time of year, and stock your refrigerator and cupboards appropriately with low-calorie foods. Also, increase your exercise in the autumn. If you are a runner, for example, and put in an average of fifteen miles a week, this would be a good time to boost that weekly average to twenty miles or more. If you do not exercise at all, autumn is a good time to get started!

Remember, you will have to go on a minidiet in the autumn *just to keep your weight from increasing*. And if you plan to start a more stringent diet at this time of the year, expect it to take a bit more effort.

HOW *WHEN* YOU EAT AFFECTS WHAT YOU WEIGH

Breakfast like a king, lunch like a prince, supper like a pauper.

It's an old adage—and one that chronobiologists have found is wise advice for people trying to lose weight. Studies have shown that calories consumed early in the day are less likely to turn into body fat than those consumed late in the day.

In one study, seven people were regimented to a single 2,000-calorie meal per day for two weeks. During the first week, they ate the meal at 7:00 A.M. (breakfast time). During the second week, the meal was switched to 5:30 P.M. (dinner time). *Everyone* lost weight—about 1¼ pounds each—during the first week; all but one person gained weight—about 1 pound each—during the second week.

This may not seem like a lot of weight. But consider how it would add up over several months. Participants in the study were losing weight at the rate of five pounds a month—or *sixty pounds a year*—on the breakfast-only diet!

Of course, eating only one meal a day is not practical—especially when you consider the ninety-minute hunger pains that urge us to eat throughout the day. (More about those in a minute.) But you can control or lose weight by taking in most of your calories early in the day—before, say, 2:00 P.M.—without sacrificing your three-meals-a-day schedule.

For many people, this means a major shift in eating habits. Most of us tend to eat more at dinner than we do at breakfast or lunch. In fact, we may skip breakfast and lunch altogether. We also tend to nibble on calorie-rich snacks during the evening and late night hours, especially while sprawled in front of the television set.

These habits spell doom if you are watching your weight. Shift those calories around. Eat your biggest meal at breakfast, your next biggest at lunch, and your smallest at dinner. And cut out those after-dinner snacks! If you get hungry in the evening, choose a low-calorie snack, such as unbuttered popcorn or fresh fruit. Save your fattening snacks (if you must have them) for the morning.

Remember, you can lose weight *without cutting back on calories* just by rearranging *when* you take in those calories. Of course, if you want to continue losing weight, you will have to cut back on your total calories, as well. But rearranging the size of your meals is a relatively simple and painless way to get your diet off to a quick start.

Weight-Watchers' Warning!

• *Beware of taking in extra calories in the autumn.* They turn into fat more easily than those consumed at other times of the year.

• *Beware of big dinners and late night snacks.* Calories consumed late in the day tend to turn into fat more easily than those consumed during the morning and early afternoon.

THE NINETY-MINUTE HUNGER CYCLES

As we discussed in chapters 1 and 2, you have ninety-minute cycles that affect the way you function, from your ability to concentrate on a project to your need to go to the bathroom. One of these rhythms is the need for oral satisfaction. About every ninety minutes, most people experience an urge to put something into their mouths. They may light a cigarette, bite their nails, or nibble on the end of a pen.

If you are on a diet, this ninety-minute urge can be dangerous, for it may send you on repeated trips to the refrigerator or candy machine.

Compounding the problem is the fact that you also have *real* hunger pains triggered by contractions of the stomach about every ninety minutes. These contractions may or may not occur at the same time as your oral urges. Either way, however, you are still fighting a continual battle against breaking your diet.

What can you do? Be aware of these cycles. Track them for a few days so you know when you are most vulnerable. (See box on page 180.) Keep low-calorie options available—water, rice cakes, fresh fruit—so when the urge to eat becomes overwhelming, you can satisfy it with the least amount of damage.

If you have the willpower, you may be able to outlast the oral urge without giving in to it. Generally, it will pass in about fifteen minutes.

Tracking Your Urge to Eat

To find the ups and downs of your daily hunger cycle, simply make a note each time during the day that you either eat or drink something or *feel* like eating or drinking something. Plot the exact time of each urge to eat on the chart shown on page 180. For best results, continue the test for a minimum of three days.

After several days of noting your hunger pains, you should begin to see a pattern develop. This should give you a good idea of when during the day you are most likely to go scavenging for food. If you are on a diet, you can then prepare yourself for those peak times with an arsenal of low-calorie foods. Sometimes, just being aware that you are entering a temporary hunger period is enough to see you through it unscathed—or, in this case, unfed.

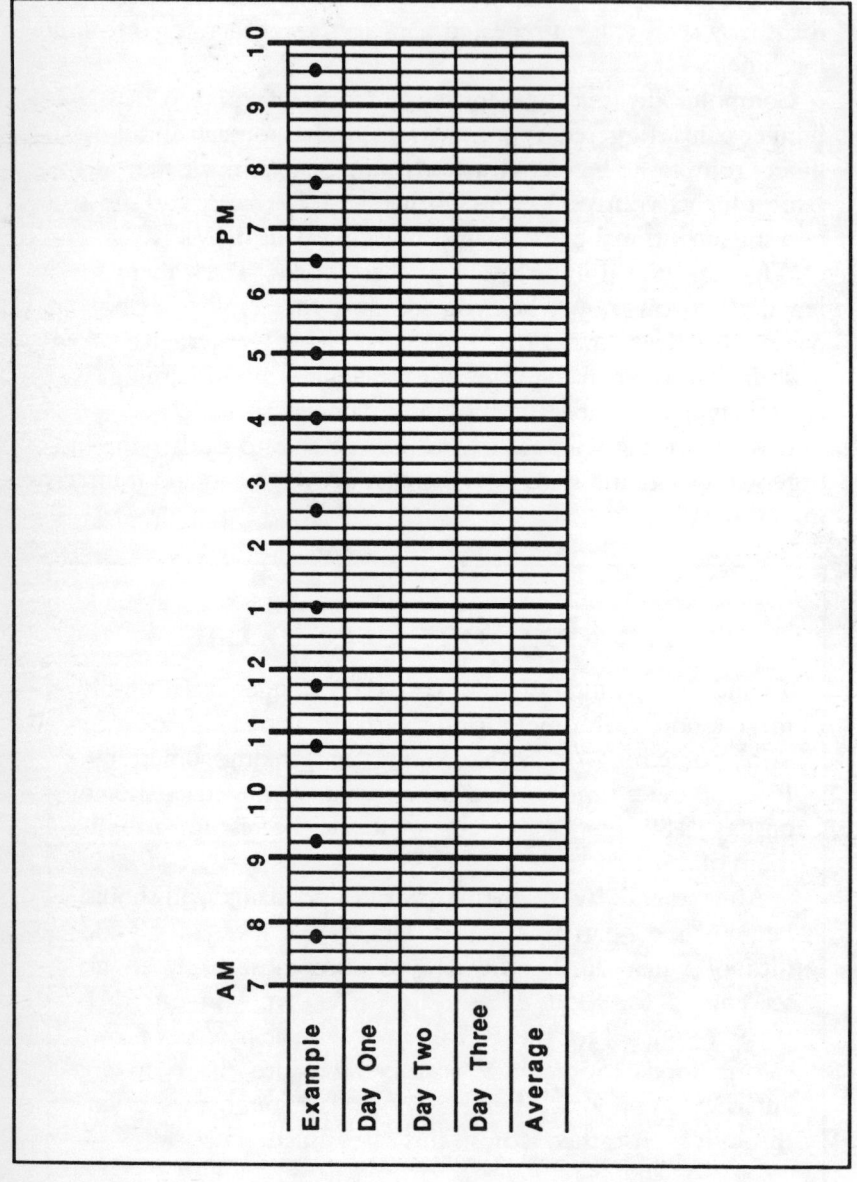

How Your Taste Buds Change During the Day

Although you are better able to discriminate between different foods in the morning, your *overall* ability to taste is more acute in the evening. That's why dinner usually seems more satisfying to us than breakfast.

That is also why you can go easier on both salt and sugar at the dinner table. A small amount will seem like more. This is good news for people who need to cut down on both of these foodstuffs for health or weight reasons (and that includes most of us).

HOW CAFFEINE AFFECTS YOUR RHYTHMS

When she went away to college, Mary got in the habit of drinking two or three cans of diet cola each night while she studied. The cola seemed to energize her, helping her to stay alert and concentrate on her books.

Soon after getting into this cola habit, however, Mary began to notice a change in her sleep patterns. Not only did she have trouble falling asleep, but she also started to wake up frequently during the night. As a result, she often arose in the morning feeling tired and unrested.

The problem? The caffeine in her diet cola. Each 12-ounce can contained about 60 milligrams of caffeine—or only slightly less than the 100 milligrams in an average cup of coffee.

Caffeine is a powerful stimulant that can have a strong effect on our internal rhythms. It can play havoc with our sleep/wake cycle, keeping us awake when we would normally be sleeping. It can also alter our alertness rhythms, speeding up and intensifying our ability to think and react.

Caffeine also raises our basic metabolic rate, which means we tend to burn off more calories after consuming it. *But,* it also triggers a release of insulin in our bodies. This, in turn, causes a drop in blood sugar, which then sets off hunger pains. So, although coffee can help you lose weight, it can also help you gain it.

Although most people are aware that caffeine is found in coffee and tea, fewer people realize it is also found in large amounts in many cola-type soft drinks. Indeed, one new cola on the market is promoted specifically for its high caffeine content. In addition, caffeine hides out in cocoa and chocolate, although in much smaller quantities. It is also an ingredient in many over-the-counter medications. (See box on page 184.)

Some studies have implicated caffeine in a variety of illnesses, from cancer to heart attacks to fibrocystic breast disease. In each case, however, subsequent studies have cast doubt on the connection between caffeine and the illness. Nonetheless, if you are pregnant, you should play it extra safe and avoid all caffeine-containing substances. Although it has not yet been proven, a connection may exist between caffeine and an increased risk of miscarriage, stillbirth, and birth defects.

The effects of caffeine—increased alertness, clearness of thought, swifter reaction time—generally show up thirty to sixty minutes after it is consumed and last for several hours. Three and a half hours after drinking a cup of coffee, for example, half of the caffeine from the coffee still remains in your body.

Once caffeine's effects begin to wear off, however, you will notice a dramatic drop in your alertness levels—a drop that is much greater than any that would naturally occur at that time of day. A morning cup of coffee, therefore, can make you more alert than usual during the morning hours, but it can also make you sleepier than usual during the afternoon.

In addition, not all of us benefit equally from the increased stimulation of a morning intake of caffeine. Much depends on whether you are a morning person or a night person. According to recent studies conducted at Northwestern University, both morning and night people perform simple mental tasks better during the morning hours after consuming the caffeine equivalent of one to three cups of coffee. But when the tasks become more complex, only night people do better under the influence of caffeine. For morning people, the higher the dose of caffeine, the more mistakes they make.

Why? Scientists speculate that it is because morning people are already more aroused and alert in the morning and that caffeine overstimulates them, interfering with their reasoning ability. Night people, on the other hand, are helped by caffeine because, without it, they would take much longer to wake up mentally in the morning.

If you are a morning person, however, you may benefit from an *afternoon* swig or two of caffeine. For, as we saw in chapter 2, your complex thinking takes a nose dive late in the afternoon. A little bit of caffeine may actually be the boost you need to keep your thinking abilities closer to their morning levels.

Large doses of caffeine—say, several cups of coffee in one sitting or ten or more cups throughout a day—should be avoided, even by night people. Too much caffeine can bring on the caffeine jitters, which are characterized by a host of undesirable symptoms, such as irregular or rapid heartbeat, upset stomach, increased blood pressure and body temperature, dizziness, headache, nausea, nervousness, insomnia, and all-around irritability.

You should also avoid caffeine after 6:00 P.M.—unless, of course, you want to stay awake and alert long past your usual bedtime. A cup of hot cocoa or tea may seem like the perfect before-bedtime drink to ease you into peaceful slumber, but the caffeine contained in these beverages will disrupt your sleep.

If you are a heavy caffeine user and decide to cut back on the drug, do it gradually—about one less cup of coffee or can of cola per day. Your biological rhythms need time to readjust. Going cold turkey on caffeine can cause headaches, drowsiness, depression, lethargy, and even nausea and vomiting. The symptoms usually begin twelve to sixteen hours after the last dose of caffeine—which is why many people fail to connect the two. In fact, those notorious weekend headaches that some people suffer may actually be the symptoms of sudden caffeine withdrawal.

Common Sources of Caffeine

Coffee (6-ounce cup)
Brewed	100–150 mg.
Instant	80–100 mg.
Decaffeinated	2–4 mg.
Coffee-grain blends	15–40 mg.

Tea (6-ounce cup)
Leaf tea	60–75 mg.
Instant	30 mg.

Cocoa (6-ounce cup)   5–10 mg.

Chocolate (1-ounce)
Milk chocolate	6 mg.
Plain	20 mg.
Baking	35 mg.

Cola-type soft drinks (12 ounces)   30–60 mg.

Over-the-counter drugs (per tablet)
Anacin	32 mg.
Cope	32 mg.
Dristan	16 mg.
Empirin Compound	32 mg.
Midol	32 mg.
NoDoz	100 mg.
PreMens	66 mg.
Vivarin	200 mg.

Caffeine Timing Tips

A well-timed dose of caffeine (in coffee, tea, cola, or cocoa) *can* help you maintain a high level of mental alertness throughout the day. The key, as in all things, is moderation. Too much caffeine will do more harm than good, throwing your rhythms off kilter and leading to the caffeine jitters.

Here are some general rules to follow:

If you are a morning person:

- Avoid caffeine in the morning or limit it to the equivalent of about half a cup of coffee (50 mg.).
- Consider having a cup of coffee or tea with lunch, and perhaps a second cup in mid- to late afternoon.
- Avoid caffeine in all forms after 6:00 P.M.

If you are a night person:

- Consider having one or two cups of coffee or tea in the morning to help you get going.
- You may wish to have another cup at lunch to help you over the early afternoon slump.
- Avoid caffeine in all forms after 6:00 P.M.

Remember: Some people are more sensitive to caffeine than others. Only you can determine, through trial and error, exactly how much caffeine helps and how much hinders your daily performance.

HOW ALCOHOL AFFECTS YOUR RHYTHMS

Alcohol *will* help you fall asleep, but, like caffeine, it will make your sleep rhythms more fragmented and less restful. If you want to wake up feeling refreshed, alert, and at your mental best, *don't* drink the evening before.

You should also be aware that alcohol hits you hardest when ingested early in the day. A single cocktail at lunch can have the same effect on you as two cocktails at dinner. That is not an excuse to drink more after twilight, however. Remember that your body is winding down in the evening, so your reflexes and thinking skills are already on the downward skid. Alcohol will only hasten that perilous descent. It will also disrupt your sleep patterns, causing you to spend more time dreaming and less time in the restorative deep sleep stage of your nightly cycle—one reason alcoholics have more nightmares than other people.

Alcohol also affects women most strongly during the premenstrual phase of their monthly cycle. A drink taken during the few days before menstruation will make you more tipsy than a similar drink taken at another time of the month. It will also have a more dramatic effect on your mood, making you more despondent or demonstrative, depending on how alcohol normally affects you.

USING FOOD TO ENHANCE YOUR DAILY RHYTHMS

A cup of caffeinated coffee isn't the only food substance that can help you overcome the early afternoon letdown and other biological slumps of your day. Some common foods, when eaten at carefully timed intervals, can help you maintain a high level of mental and physical energy throughout the day.

Conversely, you can use carefully selected foods to calm and relax you at those times of day when your rhythms are naturally slowing down, but when stress or outside events are making it difficult to unwind.

Sound like nutritional voodoo? It's not. Scientists at the Massachusetts Institute of Technology, Harvard University, and the University of London have recently discovered fascinating new information about the relationship between food and the brain.

Here, in simplest terms, is they found:

• *Carbohydrates help calm and focus your mind.* The reason: Carbohydrates stimulate the secretion of insulin in the bloodstream. The insulin then washes all amino acids except *tryptophan* from the blood. Without competition from the other amino acids, tryptophan is able to reach the brain in larger quantities, where it triggers the production of *serotonin,* the brain's calming chemical.

• *Protein foods increase your alertness and help you feel more energetic.* The reason: Protein foods introduce the amino acid *tyrosine* into the body. Tyrosine, in turn, stimulates the production of *dopamine* and *norepinephrine,* which are the brain's alertness chemicals. The effect from eating protein-rich food, however, is not as dramatic as that from eating high-carbohydrate meals.

So all you have to do is have a hamburger when you want to feel mentally energized and chocolate when you want to feel calmer, right? Almost, but not quite. There are a few other mitigating factors you must keep in mind:

• *Eating protein will not boost your mental energies if your brain already has sufficient quantities of dopamine and norepinephrine.* In other words, there is a limit to the amount of these alertness chemicals that can appear in your brain at one time. That's why eating a high-protein breakfast will not make you extra alert during the morning hours—because your alertness chemicals are already naturally on the rise then. (As you may remember from chapter 2, this is true even for night people.)

• *Eating a protein-rich food will not boost your mental energies if that food is also high in fat.* Fatty foods take longer to digest than other foods, which causes more blood to be drained from the brain to the stomach. That can slow down your thinking and make you feel sluggish. Fats also slow the absorption of protein into the body, which means that the protein's energizing effects take longer to get started.

• *Protein can be eaten either alone or with a carbohydrate food to energize the mind. For carbohydrates to have their calming effect, they must be eaten alone.* When protein and carbohydrates enter your body at the same time, the protein inevitably gets its way and makes more tyrosine available to the brain. Carbohydrates must be eaten alone if you want them to put you in a more relaxed frame of mind.

• *More is not better—just more fattening!* You need to eat only
three to four ounces of protein for it to boost your brain power
and only one to one-and-a-half ounces of carbohydrates to feel
their quieting effect. Eating more than these amounts will not
make you more alert or more calm—but it might make you fatter.
The only exceptions to this rule are people who are more than 20
percent overweight and women who are in the premenstrual phase
of their monthly cycle. Both groups may need a little more
carbohydrates to experience the calming effect.

Putting all this information to practical use may seem a bit
imposing at first. But it is really quite simple. By carefully selecting
and timing the foods you eat, you can enhance or overcome the
ups and downs of your daily alertness rhythms.

Getting Off to a Good Start

Unfortunately, as we discussed earlier, eating a high-protein breakfast will not have much effect on your alertness level because
dopamine and norepinephrine, the brain's two alertness chemicals,
are already on the rise at that time of day. Avoid high-fat foods at
breakfast, because as noted, they tend to make you sag in the
morning.

A hearty, low-fat meal (such as cooked oatmeal with raisins,
cinnamon, and low-fat milk, or whole wheat pancakes topped with
sliced fruit and cinnamon) soon after arising from bed is a good
zeitgeber, sending messages throughout your body that another
day has started. Two long-term studies—one done at the University
of California at Los Angeles and the other at Johns Hopkins
University—have found that people who eat a substantial breakfast
are likely to live longer. So don't skip it! And, remember, morning
calories aren't as fattening as evening ones!

Fighting the Early Afternoon Slump

Eating protein at lunch is a good way of fighting the early
afternoon slump, that time of day when your alertness cycle takes

a dramatic downward turn. For best results, eat the protein *first,* before you begin munching on a carbohydrate. In a restaurant, that means foregoing the bread or rolls until your cottage cheese or chicken salad arrives. At home, it means not reaching for a sweet before you have had the main course!

Avoid high-fat foods at lunch, for, as you know, they will only make you feel sleepier as the afternoon progresses. A big lunch will also have the same effect.

Fighting the Late Afternoon Doldrums

As we discussed earlier, the mental powers of morning people take a more precipitous drop during the afternoon than those of night people. If this is the case with you, try a bit of protein (or, if you prefer, a cup of coffee or tea) to keep alert and awake during this droopy part of your day.

Fighting the Late Afternoon Fidgets

Some people become fidgety or restless in the late afternoon and lose their ability to concentrate. Scientists are not sure why, but they believe it may be because these people experience a late afternoon dip in serotonin levels in the brain. In any event, the restlessness can be helped in most cases by snacking on a high-carbohydrate food.

Keeping Alert at Night

If you have an active evening planned—a concert, a get-together with friends or family, or perhaps something not quite as pleasant, such as balancing your chequebook or writing an essay—you should be careful about what you eat at dinner.

Be sure the meal is high in protein and low in fat. Eat small portions—too much food can make you sleepy.

For an extra boost, try an after-dinner cup of coffee. But remember: The caffeine can stay with you for as long as six hours. Drink it *only* if you plan on delaying your regular bedtime.

Foods to Help You Unwind in the Evening

If your goal in the evening is to relax and unwind, avoid protein at dinner and help yourself to carbohydrates instead. But don't overdo it! Although large meals can give you that snug, satisfied, sleepy feeling, they also usually pack a lot of calories—and calories, remember, turn more easily into fat late in the day.

A big evening meal can also disrupt your sleep cycle by keeping your digestive system working overtime.

Caffeinated beverages, of course, should be avoided entirely if you want a low-key, low-energy evening. Try herbal teas instead, or a grain-based beverage, such as Aromalt or Barleycup. Most supermarkets now carry these products.

Foods to Help You Sleep

You should always try to let your natural rhythms sweep you off to sleep. However, if you have trouble letting your mind drift, try a

---

Foods That Energize
(Proteins)

When it comes to energizing your mind, not all protein-rich foods are alike. The best choices are those that are high in protein *and* low in fat. Here are your best bets:

Very lean beef with no visible fat
Chicken (without its high-fat skin)
Fish
Low-fat cottage cheese
Low-fat yogurt
Skimmed or low-fat milk
Pulses (peas and beans)
Tofu, tempeh, and other soybean-based foods

carbohydrate food as a nightcap. The serotonin released in the brain by carbohydrates is believed to induce sleep.

The traditional glass of warm milk, by the way, is *not* a good nightcap, because of its high protein content. You would do better to nibble on a few crackers or a piece of toast with a dab of jelly or jam. And, of course, *no* cocoa, tea, or coffee.

Foods That Calm
(*Carbohydrates*)

Both kinds of carbohydrates—starches and sugars—can ease feelings of anxiety and frustration. However, you would be wise to choose a starchy food over a sugary one, because sugary ones are also usually high in fat. The fat adds to the calming effect of the food—but it also adds to your daily calorie count and to an increased risk of heart disease, cancer, and other illnesses.

With that warning in mind, here are your best bets for carbohydrate foods that calm. You will notice that fruits are absent from the list. Unfortunately, the natural sugar in fruit takes too long to convert into a form that will cause insulin to rise in the blood.

Starches	Sugars
Bread	Cake
Corn	Sweets
Crackers	Biscuits
Pasta	Ice cream
Potatoes	Jams and jellies
Rice	Fruit pie

Tips for a "Rhythmic" Diet

• Eat smaller meals as the day progresses.
• Avoid high-calorie snacks in the evening.
• Make an extra effort to count calories during the autumn and winter months.
• Eat proteins when you need a little boost of mental energy.
• Eat carbohydrates when you want to calm down mentally.
• Avoid fatty foods when you need to be mentally alert.
• Limit your caffeine intake to the equivalent of three cups of coffee per day (about 300 mg.) and pace it appropriately during the day.

Eight

~

Rhythms Off Beat: Jet Lag

He has called together legislative bodies
at places unusual, uncomfortable and distant . . .
for the sole purpose of fatiguing them into
compliance with his measures. . . .
—DECLARATION OF INDEPENDENCE
July 4, 1776

It was Ann's first holiday in Europe, a three-week dream vacation that had been planned to the last detail. The whirlwind tour included sites she had wanted to see for years—Big Ben in London, the Leaning Tower in Pisa, the Coliseum in Rome—and much more. Everything should have been perfect.

But after arriving in London on the first leg of her journey, Ann felt cranky and out of sorts. Her friends had warned her about adjusting to a new time zone, but this was more debilitating than the tiredness and disorientation they had spoken about. She felt downright awful. The feelings did not disappear with a nap—or even with the passage of a few more days in her new surroundings. In fact, on day three of her trip, the symptoms became worse. It was nearly impossible to sleep, especially when she found herself waking up to go to the bathroom several times a night. She felt weak, tired, even dizzy. She was unusually forgetful, constantly misplacing her room keys. And she was not hungry at mealtimes. Most annoying were the roaring headaches and recurring upset stomach that ruined several sightseeing tours. It seemed as if she would never get in sync with her surroundings.

What went wrong? Ann was the victim of jet lag, one of the most notorious and extreme examples of what can happen when the body's natural cycles are disrupted. Traveling rapidly to new time zones throws the carefully set timing of our biological rhythms into chaos, bringing on a host of maladies that can last for weeks.

Although almost everyone who flies across more than three of the earth's twenty-four time zones is affected by jet lag, not all are affected equally. Much depends on such things as your age, personality, sleeping habits, and whether you are a morning or an evening person. This chapter will look at the biological causes of jet lag—and what you can do to ease its symptoms on your next pleasure or business trip.

Your Susceptibility to Jet Lag: A Quiz

• Are you an introverted, shy person who likes to be alone?

• Are you anxious or easily stressed?

• Do you regularly use drugs, including alcohol?

• Do you smoke?

• Is your weight normal or less than normal for your age, size, and sex?

• Do you have any chronic illnesses?

• Do you routinely fly across several time zones more than once or twice a month?

If you answered yes to any of these questions, you may be particularly prone to suffer the ill effects of jet lag.

WHAT IS JET LAG?

Jet lag, simply put, is the disruption of your internal rhythms caused by a sudden change in zeitgebers—the cues from the outside world that help regulate your internal body rhythms. If you fly from New York to London on a typical overnight flight, for example, you will arrive at just about the time people in England are sitting down to breakfast. For your body, however, it is still three in the morning. That's what causes the confusion. The sun rising over London, as well as the sights, sounds, and smells of Londoners eating breakfast and going off to work, send a strong message to your inner biological rhythms that it is morning and time to get into action. But those rhythms, set for a time when normally you would be soundly sleeping, cannot readjust immediately. The switch may take days, or even weeks. The rhythms that influence your alertness and concentration, for example, can take anywhere from two days to two weeks to recover. Your heart rate can take five days to resynchronize. And urinary output, which normally is low at night when you are sleeping, can take up to ten days to readjust to your new schedule.

By the simple act of stepping off a jet plane, you have suddenly thrown all your biological rhythms into disarray. Because some cycles take longer than others to readjust, your body is not only out of sync with the outside world, it is also out of sync with itself. It is this "internal dissociation" that causes many of the physical ailments of jet lag.

WHAT CAUSES JET LAG?

Despite almost thirty-five years of studying jet lag, scientists still do not understand much about how it disrupts our rhythms, why our bodies react the way they do, and how we recover to get back in sync with the world around us. But they do know that the following factors seem to contribute to the syndrome.

- *How far you travel.* The more of the earth's twenty-four time zones you cross during a flight, the more you will suffer from jet lag. Jet lag really becomes noticeable for most people after crossing three or more time zones, although there is some evidence that crossing even one zone causes minor effects. While there are no hard rules, you should count on it taking you about one day to recover for every time zone crossed.

• *The direction you travel.* Flying east causes more severe jet lag than flying west. In fact, it can take up to 50 percent longer to recover from jet lag following an eastward flight than after a westward flight of the same distance. The reason lies with the body's natural desire to follow a sleep/wake cycle of about twenty-five hours, rather than the twenty-four-hour cycle of the sun (see chapter 1). This tendency of the body to extend the day fits nicely with a westward flight, which does exactly that. Flying to the east, on the other hand, shortens the day and is much more disruptive to your natural cycles.

Interestingly, flying north or south, no matter how far or how fast, will not cause jet lag as long as you stay close to your home time zone. For example, if you fly from Washington, D.C., to Santiago, Chile, a distance of 5,000 miles but with only one hour difference in time, you may very well be fatigued at the end of the journey, but you won't be suffering from the more debilitating effects of jet lag.

• *Losing sleep at 35,000 feet.* It is the rare person who can sleep soundly on an airplane. The loss of sleep on a long flight can severely disrupt your sleep cycles and make your jet lag that much worse. The problem is greater flying east than west, especially on overnight flights, because you are cutting hours off your normal "night" as well as trying to sleep under difficult circumstances. Most travelers who step off a United States-to-Europe overnight flight have not only lost their normal time cues, they've also lost most of a night's sleep as well.

• *Flight-related fatigue.* Everyone who flies long distances, even on north-south flights, suffers from the fatigue that comes with hours of sitting in the rarefied atmosphere of a high-flying jet. The fatigue is different, and much worse, than the tiredness that results from sitting in a car for the same amount of time. The humidity in an airplane cabin is low, making it easy to become dehydrated. The oxygen level is thin, about what you would normally experience at the 7,000- or 8,000-foot level of a mountain, and the cabin pressure isn't much better. These factors put a tremendous strain on your body that exacerbates the effects of jet lag.

It Makes No Difference Whether You Are Coming or Going

Scientists used to believe that people were more affected by jet lag after trips that took them away from home rather than after trips that returned them home. It was theorized that the established routines of home life helped travelers resynchronize body rhythms faster. However, new research indicates that the "outbound" and "homebound" differences in jet lag are really not that great, if they exist at all. Coming or going, you will still suffer from jet lag.

HOW JET LAG AFFECTS PERFORMANCE

Whether you are a business executive making a crucial presentation to a prospective foreign customer or a newly arrived tourist trying to figure out the complexities of New York's subway system, jet lag will lessen your abilities to perform mental and physical tasks. You just won't be able to think or act as precisely as you would under normal conditions. Mentally, your powers to reason logically and learn new information will diminish, as will your powers of memory. In addition, hand-eye coordination will be off, and purely physical attributes such as your strength and how fast you can run, will decline markedly.

Why? One major reason has to do with sleep. As we saw in

chapter 3, our bodies have a predetermined need for sleep, especially deep sleep and dream sleep. Deep sleep is the time when the body restores itself; it is also thought to be important for our mental health. Dream sleep is the period when our brain undergoes its daily housecleaning, sorting out the day's events and filing them into our memory. When we get too little of these two types of sleep, we feel sluggish and irritable, and our mental and physical skills decline.

Jet lag almost guarantees a long-distance traveler of at least one or two nights of fitful sleep—and thus less deep sleep and dream sleep than normal. The end result: a physical and mental dullness that can linger for days.

You sleep fitfully after a long-distance trip because your sleep cycles and your body temperature are no longer synchronized. Normally, you go to sleep while your body temperature is on the decline and awaken soon after it begins to increase. Both the length and quality of your sleep depend on this close relationship with body temperature.

Long jet flights, especially overnight flights to the east, abruptly disrupt your normal sleep time but don't immediately change the cycle of your body temperature. So when you finally do go to sleep, you are out of sync with the daily rhythms of your body temperature and, as a result, you sleep more lightly than usual. Instead of waking up restored, you wake up feeling mentally muddled.

It is not only having your body temperature and sleep cycles off balance that slows you down after a long-distance flight. Even if you were to keep those cycles in sync, you would still find your thinking powers lessened. This is because all your other cycles—from the ups and downs of blood pressure to the ins and outs of digestion—are also out of sync with each other.

This combination of things gone awry—the shallow sleeping and the total internal desynchronization of body rhythms—is what gives jet lag its one-two punch.

In one study, airline pilots were tested for reaction time and coordination one day before and one day after flying east across

Using Jet Lag to Your Business Advantage

The slowness with which your body readjusts its cycles is, for the most part, a major annoyance during long-distance travel. But a smart business executive or diplomat flying in the right direction can actually take advantage of being out of sync with the people he or she must deal with.

On a trip from Europe to Japan, for example, you should find yourself (if you got enough sleep during your flight over) more alert than your Japanese colleagues during the late afternoon and early evening of your first day in their country. The reason is simple. Many of your body's most important cycles will still be running on their "old" schedule and will peak during the Japanese evening, thinking it is late morning or early afternoon. But the cycles of your Japanese counterparts will already have peaked and started to decline. As a result, you should have more energy and be thinking faster and more clearly than your Japanese colleagues.

This advantage, of course, soon wears off as your body desynchronizes and begins adjusting to the new time. However, you may wish to catch whomever you are doing business with "off peak" and schedule a meeting the first, or perhaps even the second, evening in your new surroundings.

But there's a flip side to this coin. When flying long distances to the west, your cycles may peak at a time that puts you at a decided disadvantage. After a westward flight from Europe to the United States, for instance, you will perform poorly in the American afternoon and evening because your home time internal clock will be set for the middle of the night. Some solace, however, can be taken from the fact that recovery from jet lag is much quicker after a westward flight, so you won't have to wait as long to get into sync with your surroundings.

six times zones; their performance on the tests declined 3 to 4 percent after the long-distance flight. The test results were not quite as bad after a similar westward flight. In another study, a group of pilots were asked to operate airplane simulators before and after flying eastward across a minimum of six time zones; the pilots operated the simulators less efficiently for as long as five days after arriving in the new time zone.

For the casual traveler, this weakening of mental and physical abilities may go unnoticed or may cause only a minor annoyance. But for travelers conducting important business or negotiations, for world-class athletes competing in international competitions, and, obviously, for pilots responsible for the lives of their passengers, even a small loss of physical or mental stamina can be critical.

COPING WITH JET LAG

Travelers have come up with almost every imaginable home remedy for jet lag, ranging from aviator Wiley Post's practice of radically disrupting his sleep patterns for weeks before a flight (see box on page 203) to the use of acupuncture to relieve symptoms after a flight.

Scientists, too, have joined the search for a quick, easy way to escape the effects of jet lag, but progress has been slow. Elaborate anti–jet lag programs have been developed, but they are all complicated and time-consuming, and each has its critics.

The most popular anti–jet lag program—one, in fact, used by President Reagan—was developed by Dr. Charles Ehret, of the Argonne National Laboratory. Ehret's approach to beating jet lag involves manipulating light, diet, caffeine intake, and sleep (see pages 209–210). The State Department uses a simpler program, developed by the Walter Reed Army Institute of Research in Washington, D.C.

Some researchers have begun to study the ability of various drugs to reset the body's biological clocks. Indeed, one biologist was so excited by the effect of a sleep-inducing drug on the cycles of golden hamsters that he raised the possibility of a jet lag pill being just over the horizon. However, other scientists doubt that overcoming jet lag will ever be as simple as taking a pill.

Yet, while no sure cure exists for jet lag, there are several things you can do to lessen its effects. First, though, you must understand that there are some aspects of jet lag over which you have no control. You can't do much about the direction of your flight, for example; nor can you suddenly change your basic personality or your weight, which also play a role in jet lag.

Knowing these uncontrollable factors, as well as the controllable ones, will allow you to gauge your vulnerability to jet lag and decide how much effort you want or need to put into lessening its effect on your next trip.

BEYOND YOUR CONTROL

• *Direction of flight.* As stated earlier, flying eastward causes more problems than flying westward, and it takes significantly longer for the body to recover. Indeed, scientists have measured the speed with which some cycles reset and have found that they readjust at an average rate of forty-five minutes per day following a long westward flight and only thirty-six minutes per day after a similar flight to the east.

• *Age.* People of all ages, including infants, suffer from jet lag. Generally, however, the older you are, the worse the effects of jet lag. Scientists believe this is due to people becoming more suscep-tible to sleep disturbances as they age—a key part of jet lag. Studies also indicate that our cycles become more resistant to change as we age, making them less able to adapt quickly to a new time zone.

• *Weight.* Overweight people tend to be less bothered than thin people by having their normal mealtimes disrupted. Thus, over-weight people are less likely to suffer hunger pang attacks at odd hours when they travel. The advantage is only slight, however, and hardly worth fattening up for.

• *Personality type.* People who like to socialize, who get out of their rooms and mix with people, adjust more quickly to a new time zone than those who stay in their hotels by themselves. By staying indoors and away from other people, you are isolating yourself from many of the time cues your body needs to reset its biological clocks. The result is that they take much longer to adjust to the new time zone. Extroverts also seem to have a natural biological advantage over introverts because their daily temperature cycles tend to be more variable than the cycles of introverts, making adaptation to a new time zone that much easier.

• *Owls versus larks.* In trying to cope with jet lag, the advantage of being either a lark (morning person) or an owl (night person) depends greatly on which direction you are traveling. Larks, with body temperatures that peak earlier in the day than owls, have an advantage on eastward flights because the trip is advancing the hours, which is in tune with a lark's naturally "phase advanced" rhythms. Owls have the advantage on westward trips, which lengthen the day, matching their "phase delayed" natures. Larks have an additional advantage, no matter which direction they travel, because they get up earlier and are exposed to more sunlight. Sunlight, remember, is one of the strongest cues your body uses to reset its cycles. But owls have their own advantage because they tend to be extroverts, which helps deal with jet lag.

• *Your fixed cycles.* As each of us is born either an owl or a lark, so, too, are we born with a certain ingrained flexibility in the adaptability of our cycles. Jet lag scientists have devised three groups into which most of us fall: inert, intermediate, and labile. Inert people, as the name implies, are those whose cycles tend to be fixed in time. They have the greatest trouble readjusting to a sudden shift in time zones and suffer jet lag to a much greater degree than other people. Labile, which means unstable or easily changed, defines people whose rhythms shift readily. They, obviously, have the least trouble with jet lag. Intermediates are those who fall in between. NASA scientists believe labile people will experience less difficulty in adjusting to the unusual sleep/wake cycles common in space flight and would be the best candidates to become astronauts. Unfortunately, there is no simple test to determine where you are on this spectrum, but it explains why two people of similar age and physical condition can have dramatically different reactions to jet lag.

Jet Lag: A Wiley Post Discovery

On a wet, gray morning in the summer of 1931, pioneer aviator Wiley Post pushed the throttle forward on his single-engine propeller airplane, the *Winnie Mae,* and lifted off from a grass airfield in New York City. During the next eight days, Post circled the globe, a feat that stunned the world because of its audacity.

Years later, scientists became interested in the flight, not so much for its daring, but for Post's "special homemade course in physical and mental training." Post was the first person to recognize the effects of jet lag and the danger they present to pilots. His training course was designed to counteract those effects—and it was developed more than a decade before the first jet took to the skies.

Post, in outlining his anti–jet lag program, was aware that the constant changes in time zones during his around-the-world flight would "bring on acute fatigue if I were used to regular hours. So, for the greater part of the winter before the flight, I never slept during the same hours on any two days in the same week. Breaking oneself of such common habits as regular sleeping hours is far more difficult than flying an airplane." Post also changed his diet and discovered that if he ate less, he needed less sleep.

By repeatedly mixing up his time cues, Post was putting his body in a labile—or flexible—state that may well have helped as he flew through twenty-four time zones in a week and a day. For most travelers, however, spending a winter following an erratic sleep schedule is neither practicable nor desirable. Fortunately, scientists now know of less disruptive measures to combat jet lag.

THINGS YOU CAN CHANGE

Once you have an idea, based on the uncontrollable factors, of how susceptible you might be to jet lag, it's time to decide how much effort you want to put into fighting back. If you are young, healthy, active, and outgoing, you may want to look at some of the simpler steps to counteract jet lag. If, on the other hand, you think jet lag could be a serious problem, you might want to undertake a more involved program to try to avoid its worst effects.

While jet lag experts may disagree on the details of various countermeasures, most agree on the basic do's and don'ts. You can fight jet lag at three basic stages of your trip—preflight, during flight, and postflight. Each stage requires different actions on your part.

Preflight

• *Consider group travel.* If you are particularly susceptible to jet lag, you should think about traveling with a tour group. Several studies have shown that individuals who travel in groups are significantly less affected by jet lag than those who travel alone. Socializing is known to be an important time cue for biological cycles, and people who travel in groups spend a lot of time interacting with each other.

• *Eat light.* Eat lightly for about twenty-four hours prior to your flight and, if practical, try to shift the timing of meals to more closely coincide with the mealtimes of your destination. Large, heavy meals at your normal time simply reinforce your home schedule, so avoid them, if possible.

• *Think of your destination.* On the day of your flight, adjust both your watch and your thinking to the time of your destination. In the hours before boarding your airplane, avoid references to your home time zone as much as possible and focus on the schedule at your destination.

• *Shift your sleep cycle.* For a day or two before your flight, try to get a head start on changing your sleep cycle. If you are flying to the east, say, from New York to Paris, you should try to go to bed and wake up an hour or so earlier than usual to start the "phase advance" mechanism of your biological rhythms. If you are flying west, you should stay up an hour or so later at night and sleep a little longer in the morning to start "phase delaying" your system.

• *Simplify your life.* Try to be in control of your schedule for a day or two before a flight. Make sure you get plenty of rest, eat properly, and keep stress to a minimum. If you begin a flight tired and frantic, you will find it all the more difficult to combat jet lag.

In Flight

• *Forget where you've been.* Now, even more than before your flight, it is important to think in terms of where you are going. Make certain your watch is set to the time of your destination and try to key your activities to that time. One New York psychiatrist attaches so much importance to this that she uses mental imagery techniques to help people fight jet lag (clients imagine themselves in a new time zone, and this helps to overcome jet lag).

• *Time your meals.* If the airline is serving a large dinner when it is 2:00 A.M. at your destination, don't eat. If you are on a night flight from the USA to Europe, it's already well past dinner time in Europe when your plane takes off, so you can count on the in-flight meal being improperly timed. Try to eat a light meal before getting on the plane to tide you over. On westbound flights, dinner might be served when it is only the middle of the afternoon at your destination. You might want to eat lightly or, depending on your arrival time, wait until you have landed to eat a big dinner.

• *Drink lots of liquids.* Airline cabins are very dry and your body becomes dehydrated during a long flight. Drink two to three glasses of water, fruit juice, or decaffeinated coffee for every four hours of flying time.

• *Control your caffeine.* There is some controversy among jet lag experts about the use of caffeine—primarily through drinking coffee—to help control jet lag. Generally, however, coffee or tea can speed the adjustment of biological cycles if you drink them at precisely the right time. The US State Department's plan says that if several cups are consumed early in your normal day, they will delay the body's rhythms in preparation for a westward flight. If you drink them in the early evening, they will help advance the rhythms in preparation of an eastward flight. For this effect to work, however, you must avoid caffeine for a day or two prior to and following a trip. Other experts don't agree, and say that caffeine, especially if you drink it during an overnight eastward flight, will simply keep you awake when you should be sleeping. All agree, however, that drinking caffeine at several different times during a long flight is a bad idea. Caffeine is also a diuretic, which is not what you want in the dehydrating atmosphere of a jet airplane.

• *Control the sunlight.* If it is nighttime at your destination and sunlight is streaming through your airplane window, pull the blinds. Sunlight, remember, is one of the strongest time cues for the body, and you should carefully control your exposure to it.

• *Sleep at the right time.* If you are flying to the east, again using the overnight flight from the USA to Europe as an example, remember that most people at your destination are already fast asleep when you board the airplane. Get on board, settle in, and go to sleep as soon as possible. You might be tempted by the meal or the movie, but staying up may cost you a day or two more of jet lag. On the other hand, on a westbound flight you want to stay awake longer; sit back and enjoy the movie. Make sure, however, that you still get at least a few hours of solid sleep if you are flying west at night.

• *Avoid smoking and drinking alcohol.* The pressure inside the cabin of a jet flying at 30,000 feet is about the same as it would be if you were standing at the 5,000-foot level on a mountain. Both smoking and drinking alcohol interfere with the body's ability to process oxygen and can make your body feel like it is at 10,000 feet or more. The added stress is substantial and can make a significant difference in how you feel when you get off the plane. Alcohol, like caffeine, tends to dehydrate your body.

• *Keep comfort in mind.* It is difficult to sleep half sitting up in the rarefied atmosphere of a jet. It is more difficult still if your feet are cramped because you stowed a bag under the seat in front of you or if you're pinned in a window seat by the people sitting next to you. A frequent flier for NASA recommends booking an aisle seat as far forward as possible, especially on jumbo jets. You can easily get up and walk around if you are on the aisle, and there is less engine noise in the forward seats. On jumbo jets, there is also a subtle fishtailing motion toward the back of the plane, something your mind may not notice, but your muscles will.

Postflight

• *Switch your activities to the new time.* Upon arrival in the new time zone, get in step with as many of the local time cues as possible. Avoid doing activities based on your old schedule— particularly sleeping or eating.

• *Don't take a nap.* You will be tired after a long flight, if not immediately because of jet lag, then because of the strain of sitting in an airplane cabin for hours. Resist the temptation to go to your hotel room and take a nap. Although you may feel better for a few hours after the nap, it will only reinforce your old schedule and slow your adjustment to the new time zone. Staying awake that first day, on the other hand, will build up a sleep debt that will make it easier for you to fall asleep at the new, earlier bedtime. So, after your flight, take a quick shower to refresh yourself, then get out of your room and into the sights and sounds of your new surroundings.

• *Expose yourself to the sun.* The sun is the strongest zeitgeber, so walk outdoors, eat at pavement cafes, and sit in the park when you first arrive in a new time zone. Not only are these activities relaxing, but they also may speed your readjustment to the new time zone by as much as 50 percent.

In fact, US researchers at the Oregon Health Sciences University of Portland have devised a schedule of exposure to light for people to follow during the first few days of long-distance travel. It is based on the finding that exposure to light early in the day helps shift the body's rhythms forward, while exposure late in the day shifts them backward. Although still experimental, you may find the schedule useful in speeding up your readjustment to a new time zone.

Here's how it works:

If you are traveling to an earlier time zone (west)

Time Difference	Go Outside
2 hours	4 P.M. to 6 P.M.
4 hours	2 P.M. to 6 P.M.
6 hours or more	12 NOON to 6 P.M.

If you are traveling to a later time zone (east)

Time Difference	Stay Indoors	Go Outside
2 hours	–	6 A.M. to 8 A.M.
4 hours	–	6 A.M. to 10 A.M.
6 hours	–	6 A.M. to 12 NOON
8 hours	6 A.M. to 8 A.M.	8 A.M. to 2 P.M.
10 hours	6 A.M. to 10 A.M.	10 A.M. to 4 P.M.

Be sure you are outside during the appointed hours, even if it is cloudy. It's the timing, not the intensity of the sunlight that's important. Also, this schedule is based on a 6:00 A.M. sunrise and a 6:00 P.M. sunset. Readjust the times if sunlight is not available between these hours.

• *Talk to people.* Socializing is also a strong zeitgeber. Whether you are traveling alone or in a group, interact with as many people as possible. Ask directions, talk to shopkeepers, share a lunch table—do whatever you can to increase the social cues you receive. These social cues reinforce the biological cues of sunlight and darkness, and will speed the readjustment of your cycles.

• *Watch what you eat.* High-carbohydrate foods, because of the chemical changes they cause in the brain, make us sleepy. A high-carbohydrate meal—say, a pasta salad or a rice and vegetable casserole—just before bedtime may help you get to sleep. You may also want to avoid high-carbohydrate foods during the times of day when you want to stay awake. (For a list of high-carbohydrate foods, see chapter 7.)

• *Don't drink caffeinated beverages or alcohol for a day or two following your flight.* Both caffeine and alcohol can interfere with the readjustment of your cycles.

Special Tips for Athletes

If you are traveling across three or more time zones to compete in an athletic event—say, from Western to Eastern Europe to run in a marathon—you'll need to take the following extra precautions if you want to perform at your best:

• *Try to arrive several days early at the site of the event.* Traveling across three or more time zones, whether in an easterly or westerly direction, can diminish your athletic performance for as long as six days. This is especially true of sports involving dynamic muscle strength, endurance, and vigilance, such as wrestling, running, and baseball.

• *If you will be traveling across many time zones—from the United States to Japan, for example—try, if possible, to reach your destination by westward flights rather than eastward ones.* Your jet lag symptoms will be less intense, and your readjustment period will be shorter.

The Ehret Anti—Jet Lag Plan

The Ehret plan, developed by Dr. Charles Ehret of the Argonne National Laboratory in America, recommends a "feast and fast" approach to preparing your cycles for a change in time zones. Ehret believes that by eating a lot one day and then a little the next for several days before a long-distance flight, you can make your body more labile, or adaptable, to shifting time cues. He also recommends using high-protein meals for energy and high-carbohydrate meals to help you sleep.

Despite the popularity of Ehret's program, some scientists are skeptical. These scientists challenge, among other aspects of the program, the idea that you can get a quick fix of energy from eating a high-protein meal. Still, many travelers swear by the program and use it frequently.

The specific steps of Ehret's program vary slightly, depending on the length and direction of flight, but the following plan—for a five- to six-hour eastward flight—is typical:

• *Three days before the flight:* Eat at your normal time and in normal amounts, but eat a high-protein breakfast and lunch, and a high-carbohydrate dinner. Drink coffee or other beverages with caffeine only between 3:00 and 5:00 P.M.

• *Two days before the flight:* This is a "fast" day. Use the same protein/carbohydrate meal plan you did on the previous day, but limit yourself to 800 calories. Again, limit your caffeine to late afternoon.

• *Day before the flight:* Eat substantially, using the same protein/ carbohydrate meal plan as before.

• *Day of the flight:* Get out of bed slightly earlier than usual to begin advancing your sleep cycle. Use the same protein/carbohy-drate meal formula and limit yourself to 800 calories.

• *During the flight:* Sleep as much as possible, but wake up at least a half-hour before the breakfast hour of your destination. Eat as much as you want on your first day in the new time zone, using the protein/carbohydrate formula of previous days. Go to bed by 10:00 P.M.

STRICTLY BUSINESS

A businessperson planning a short two- or three-day trip overseas can take another approach to coping with the sudden change in time zones—simply ignore it.

Indeed, if you are only going to spend a brief time in a new time zone, trying to adjust to it will be both fruitless and counter-productive. Remember, most biological cycles take from several days to several weeks to readjust. So, if you spend two or three days in Japan, then fly home, you will be returning to your old routine just as your cycles are in the midst of shifting to the Japanese routine. The result: two back-to-back cases of jet lag that are doubly difficult to overcome.

To minimize this problem, Dr. R. Curtis Graeber, a NASA scientist and an expert on jet lag and international travel, recommends that business travelers make sure they get enough sleep for the meetings they must attend, then get back home as fast as possible. Don't worry about trying to make it easier for your cycles to adjust to their new environment—you won't be there long enough to make it worthwhile.

Of course, even a brief stay will cause your scores of biological cycles to begin shifting, so you can't expect to feel or perform at 100 percent of your ability. But, by getting enough sleep, you shouldn't encounter any serious physical or mental sluggishness.

To help you sleep in a different time zone, Graeber suggests the use of triazolam, a prescription drug that has long been used to treat insomnia. Triazolam—marketed in Britain as Halcion—provides a solid six hours sleep and is useful for forcing yourself to sleep. While the drug isn't physically addicting, it can cause psychological dependence with prolonged use, so it should only be taken for brief periods.

Business travelers on a three-day trip to Europe might take a small amount of the drug just as they board a night flight so they can sleep solidly on the airplane. But a word of caution: Although using the drug may allow you to sleep, it may also make you groggy and slow to respond during an in-flight emergency. Indeed, most authorities recommend specifically against taking sleeping drugs for jet lag because of this potential flight safety problem.

Whether or not you use a sleeping pill on the airplane, once you have arrived at your destination, you may wish to use a sleeping pill for a couple of nights to help you get to sleep at the local bedtime. Remember, though, the type of sleep you get under the influence of a sleeping pill is much lighter than under nondrugged conditions. (See chapter 3.) So, although you will get more deep and dream sleep than you would if you stayed awake most of the night, you won't get as much as you would if you could fall asleep naturally.

Stop using the drugs once your business is completed and you have returned home. Although jet lag will be with you on the return trip, your cycles will not have shifted significantly, and you should recover within a day or two.

Tourists and others on extended trips should avoid using sleeping pills because they will slow adjustment to the new time. By

---

Seasonal Clock Changes: Jet Lag in Your Bedroom

Even if you're not a member of the jet set and have never traveled at a speed faster than sixty-five miles an hour, you have experienced the jet lag syndrome. For each time you "spring forward" and "fall back" to make the biannual one-hour clock change, you have, in essence, suddenly crossed one time zone.

A one-hour shift doesn't seem like much, especially when your basic daily routine remains the same, but it does appear to have a dramatic impact on performance levels. Scientists have found that in the week after we lose an hour due to the spring shift in time—the same as flying eastward one time zone—traffic accidents increase sharply, as much as 10.8 percent in one study. Another study showed a 3.4 increase in accidents in the week following our autumn gain of an hour.

The time shift and the accompanying "desynchronosis" is apparently significant enough to cause motorists to become more irritable and aggressive, and less able to make critical decisions. The effects are also long-lived, with a noticeable increase in traffic accidents continuing for at least a week after the time shifts. Scientists believe the behavior change is linked to changes in our sleep pattern.

taking sleeping pills night after night, you also run the risk of creating a rebound effect, where it becomes difficult to get to sleep without the drugs. Using sleeping pills for more than a few days, combined with the many effects on your body of shifted time cues, will, according to Graeber, turn you into a "basket case."

Tips for a Journey Through Time

In the modern age of jet travel, you journey through time as well as distance. While it may take only a moment to reset your watch ahead or back five or six hours, it takes days—even weeks—for your internal rhythms to set themselves to the new time. Travelers jetting across more than a few time zones should remember the biological strain they are putting on their bodies and take steps to lessen that stress.

• Assess your susceptibility to jet lag and determine how much effort you'll need to put into countering its effects. (See chart on page 194.)

• Take the appropriate steps before, during, and after a long-distance eastward or westward trip to minimize the effects of jet lag.

• Take extra care driving a car (or doing other activities requiring high levels of coordination and alertness) during the days following the seasonal one-hour clock changes.

Nine

~

Rhythms Off Beat:
Shift Work

The night cometh, when no man can work.
—JOHN 9:4.

Bob had been working as a computer technician for only a few months, but he was already earning the respect of his supervisors. When something went wrong with the computers that controlled the company's assembly line, Bob could be counted on to pinpoint the trouble quickly and fix it. He was a fast learner and always kept up to date on new computer technology.

His only problem with his new job was the size of his paycheck. He and his wife, Jenny, planned to have a baby within a couple of years and were trying to save enough money to buy a small house. The easiest and fastest way to bring in some extra cash, Bob decided, was to switch to the night shift—11:00 P.M. to 7:00 A.M. It offered considerably better pay.

At first, Bob enjoyed the change. The pace at the plant was slower and quieter at night. It was almost like working at a different company. For the first few nights, Bob noticed that he was able to figure out some of the more complicated computer problems he had to deal with faster than usual.

But he also found that he was missing problems during his routine checks of the company's various computer systems— problems he normally would have caught. And his typing skills, which had never been great, worsened noticeably. He was constantly making mistakes when he typed commands into the computers. His productivity began to suffer.

Bob assumed things would get better after a few days—or nights—had passed, but he found that his job performance continued to deteriorate. The improved problem-solving ability he had noticed during the first couple of nights soon disappeared.

He also was constantly tired. Getting "a good day's sleep" was proving to be more difficult than he had anticipated. The curtains in his bedroom were not thick enough to make the room completely dark, and he was often awakened by a host of daytime noises—the screech of car tires on a busy road nearby, the hammering of city work crews, the roar of a passing airplane, and the shouts of his neighbor's children. Instead of his usual eight hours of sleep each night, Bob was averaging only five or six.

Jenny knew Bob was having a difficult time adapting to the night shift and tried to be understanding, but she felt resentful when Bob slept past 10:00 A.M. on the weekends. She insisted he get up at a normal hour so they could get out and do things together.

Within a few months, Bob began taking sleeping pills to help him sleep and antacids to calm his perpetually upset stomach. No matter how hard he tried to adapt to his new schedule, he couldn't shake the feeling of being "off center." He and Jenny argued almost daily.

What Bob was experiencing is common to almost all shift workers. Humans are by nature creatures of the day, and when we try instead to become creatures of the night, we pay a heavy physical and mental price.

About 25 percent of us work the night shift at least occasionally, and that number promises to increase in future years. In ancient times, it was only the herders and gatekeepers who had to work shifts. Now, given our fast-paced, consumer-oriented society, it seems that almost every profession or trade involves some shift work.

In this chapter, we'll look at how working the night shift, whether temporarily or permanently, wreaks havoc on biological rhythms. You will learn how to determine if you can handle shift work or if you'd be better off sticking to the nine-to-five tradition. If you are one of the millions of people who must work in the dead of night, then you'll learn ways to cope with the stresses and strains inherent in shift work. Despite all the negative aspects of shift work, some people actually enjoy working nights. We'll see what their secret is.

Finally, we'll look at what companies can do to change their shift schedules so they will be more in sync with the body's natural rhythms and, as a result, easier on their workers.

Are You Fit for Shift Work?
A Quiz

• Are you a morning person who goes to bed early and rises at the crack of dawn?

• Do you have trouble getting to sleep or sleeping soundly?

• Are you middle-aged or nearing retirement?

• Are you an introvert who is uncomfortable in new situations?

• Do you often suffer from indigestion or have more serious digestive system problems, such as ulcers?

• Are you married?

• Do you have young children?

• Do you hold down a part-time job in addition to your main employment?

If you answered yes to any of these questions, then you may not be suited for shift work.

SHIFT LAG: THE JET LAG
OF THE WORKING WORLD

It was the Industrial Revolution of the eighteenth and nineteenth centuries that triggered the widespread use of shift workers. Factories needed them to keep production—and profits—high. Today, shift workers are still an integral part of our factories, but that is not the only working site where they can be found. Nurses, police officers, air traffic controllers, store clerks, and many other types of workers also ply their trade through the night. In fact, as the base of our economy has shifted during the past several decades from heavy industrial manufacturing to service industries (those that produce services rather than goods), the demand for shift workers has actually increased. Hospitals, hotels, airports, and all-night convenience stores need night workers as much as factories do. Indeed, shift work has become such an accepted part of our society that as many as one out of four workers can be found doing it, at least occasionally. That number is expected to continue to increase during the next few decades.

Although shift work may be popular with employers, it is not popular with employees. Shift work takes its toll on workers, both mentally and physically. The problems caused by shift work are in many ways similar to those caused by jet lag—so similar, in fact, that the overall out-of-sorts feeling that comes with working nights has been dubbed shift lag by scientists.

Shift workers, like jet travelers, experience a disruption of their natural circadian rhythms. Changing to an eight-hour work schedule that starts at 11:00 P.M. instead of 9:00 A.M., for example, can have the same impact on your body's rhythms as crossing fourteen time zones, or flying more than halfway around the globe. Even shifting to a work schedule that begins at 3:00 P.M., which may not seem to be much of a change, is the equivalent of crossing six time zones, the distance between New York and Belgium.

By working nights, you are forcing your body to be active when it normally wants to sleep and to sleep when it normally wants to be active. Your biological systems, which are linked with the daily rhythm of the sun, are thrown into chaos by this reversal of the normal schedule.

To make matters worse, many people who work the night shift change their hours back to that of a daytime person at the weekends. Rhythms that have spent the work week trying to synchronize with the night, suddenly are confronted with a whole new set of zeitgebers and start to revert back to a daytime schedule. Then, when the weekend ends, they are faced with the night schedule again and must begin shifting back the other way. The result, of course, is permanent desynchronization and a body that is *never* in complete harmony with itself.

It is little wonder that shift workers, almost without exception, do a poorer job than their daytime counterparts. Nor should it be surprising that shift workers suffer from more stress, gastrointestinal disease, respiratory ailments, and other disorders than other people. Worse yet, shift workers seem to have weaker immune systems than day workers, and that, according to speculation among scientists, may lead to a higher death rate.

ARE YOU SUITED FOR THE NIGHT SHIFT?

No one is perfectly suited for shift work. Given the physical, mental, and social stresses that come with working nights, none of us can do it regularly without suffering some ill effects. However, some of us handle shift work much more easily than others. Scientists have identified certain characteristics that seem to help differentiate between individuals who are suited for shift work and those who are not. These traits are not foolproof, but, by measuring yourself against them, you can get a fairly good idea of your ability to handle shift work.

• *Larks versus owls.* The biggest problem for shift workers is sleep—or the lack of it. As we will discuss in detail later, shift work plays havoc with our sleep/wake cycle, and even people who permanently work the night shift suffer from a chronic lack of sleep. Most shift workers go to bed in the morning, when the sun is rising. Morning people, or larks, are naturally alert at this time of day. Thus, they find it doubly difficult to try to sleep after coming off the night shift. Evening people, on the other hand, experience a natural drowsiness early in the morning. Thus, they are better able to fall asleep after a night's work and find it easier adjusting to the sleep regime demanded by working nights.

• *Personality.* If you are outgoing, you are likely to have an easier time coping with shift work than a colleague who is an introvert. One reason is that extroverted people tend to be evening people. Also, the daily temperature cycles of extroverted (and evening) people tend to be less rigid than those of introverted people, which makes it easier for an extroverted person to adapt to a new sleeping schedule.

Another important personality factor that will influence your adaptability to shift work is how well adjusted, or stable, you are. People with stable personalities, whether they are outgoing or shy, seem to deal better with shift work than neurotic people. Scientists are trying to determine why.

The best shift worker, then, is a stable, outgoing person, while the worst is a neurotic introvert. However, if you fit into the former category and are about to go on the night shift, be aware that over the long term your unnatural work hours can actually change your personality and make you neurotic! The reasons for this personality shift are unclear, but, ironically, the changes it brings will make you no longer suitable for shift work.

• *Flexible versus rigid rhythms.* As we saw in chapter 8, people with inert, or rigid, biological rhythms have a more difficult time adjusting to a new time zone than individuals with labile, or flexible, rhythms. The same is true of making the adjustment to shift work. Studies show that people with rigid sleep patterns are more bothered by noises and other disturbances when they are trying to sleep during the day than people with more flexible rhythms. Again, evening people and extroverts tend to have more flexible rhythms than morning people and introverts.

• *Age.* Working the night shift—even if you have done it for years—becomes much more difficult as you grow older, particularly after the age of fifty. The reason for this inability to cope with shift work is unclear; however, scientists suspect four major factors: (1) the cumulative adverse effects of shift work, (2) the general weakening of health that comes with age, (3) a natural decrease in the flexibility of daily rhythms, and (4) the deterioration in sleep rhythms that is typical among older people. This last factor is the most important, for the fact that our sleep cycle becomes more fragile as we age tends to intensify the other physical problems of aging. Also, as we grow older, we tend to become morning people, even if we were owls during our youth. That gradual transformation from an owl to a lark works against aging shift workers.

• *Health.* The state of your health can also affect your adaptability to shift work. The poorer your health, the greater the difficulty you will have adjusting to shift work, because you need to be in good physical shape to withstand the rigors of a rotating schedule. Conversely, the longer you work the night shift, the more unhealthy you are likely to become. As we will discuss in detail later in this chapter, most shift workers get too little sleep, have poor diets, and often live with a high level of domestic and social stress. None of those things are good for your health.

• *Marital status.* Shift work puts a great strain on family life. When one spouse works during the night and sleeps during the day, the normal life of everyone in the family is disrupted. The shift worker is gone during the early evening hours, when family activities usually take place. Meals, an important ritual in family life, are difficult to plan. The home, usually bustling with activity in the morning, must be quiet so the shift worker can sleep. Finally, normal social life is almost nonexistent.

Because most shift workers want to have as normal a family life as possible, they often try to rearrange their schedules at the weekend so they can become day people, if only for a couple of days. This makes it all the more difficult to readjust to the night shift when Monday rolls around.

• *Men versus women*. With the exception of nurses, shift work has traditionally been done by men. Indeed, decades ago, an international organization passed a resolution banning the use of women in night work. Many countries still have such laws. Today, however, women are being called upon to do shift work in increasing numbers. Unfortunately, few studies have been done to determine if shift work affects women differently than men.

The key concern is that shift work will disrupt women's menstrual cycles. So far, studies show contradictory results and, for the moment, scientists believe women are as physically able as men to cope with shift work.

However, in a society that still sees women as the primary caregivers and caretakers of their families, women shift workers face additional stresses not encountered by men. A woman who works shifts must frequently clean house, do laundry, fix meals, and get her children off to school at normal times—in other words, when she should be sleeping. It is a burden that only adds to the natural stress of shift work.

• *Commitment*. How you view shift work is as important a factor in how well you adjust to it as any of the other categories we've discussed. If you are committed to restructuring your life to match an unusual work schedule, then adjusting to shift work may be relatively easy for you. Committing to shift work, however, is not easy. All your activities—when you sleep, when you eat, when you socialize—must be dramatically altered.

If, on the other hand, you try to remain a day person by holding on to as much of your daytime schedule as possible, any prolonged shift work will be difficult, if not impossible. Your natural rhythms, remember, are keyed in large part to sunlight, so trying to live the life of a day person while working nights is like trying to live in two time zones at once.

A study of part- and full-time night nurses showed that the full-time night nurses did not suffer the midshift sag of energy and alertness that plagued the part-time workers. Researchers found that many more of the part-time workers had children at home or were working another job during the day—factors that meant they could not fully adapt their schedules to night work.

HOW SHIFT WORK AFFECTS YOU

As Bob, the computer technician, discovered, the effects of shifting to night work are dramatic and wide-ranging. Every aspect of his life, from his job performance to his relationship with his wife, suffered when he began working nights. Bob's experience is typical of what happens to many shift workers. In this section, we'll consider in more detail the major problems you are likely to suffer if you work shifts regularly.

Working the night shift is not all doom and gloom, however. For some people, it has its advantages—and we'll see why.

How Shift Work Affects Your Social Life

If you are a young, single worker who likes to spend evenings at nightclubs or restaurants, then you'd be wise to stay away from shift work. Even if your idea of a good time involves something more mundane, such as joining a bowling league or attending a meeting of a local community organization, you still should avoid shift work. For it is nearly impossible to carry on anything even resembling a normal social life when you work the afternoon/evening shift (roughly 2:00 P.M. to 11:00 P.M.) or, to a lesser degree, the night shift (11:00 P.M. to 8:00 A.M.).

The reason for this is self-evident. Our society is oriented toward day work, so most social activities are scheduled for the evening— the time of day when shift workers are either at work or getting ready for work and, thus, unable to participate.

The disruption of your normal social life shouldn't be underestimated, for it means more than simply missing out on a few meetings or nights out with friends. Shift workers have a completely different structure to their social lives and, as a result, they tend to lead a more isolated existence than their daytime colleagues. The loss of a normal social life is so serious that it ranks with loss of sleep as one of the two most frequent complaints from shift workers.

Studies show that if you work shifts, you are less likely to join clubs and social organizations and, if you do join, you probably won't hold an office or be active in the group. Shift workers usually do not become actively involved in politics, parent-teacher associations, or even in recreational sports leagues.

Because of this social isolation, friendships are more difficult to develop. Shift workers often have fewer friends than their daytime colleagues, and the friends they do have tend to be limited to the people they work with.

For those who like solitude, however, there are definite advantages to the isolated lifestyle that comes with shift work. Many night workers take advantage of having their days free to pursue hobbies and projects with an intensity not possible for day workers. Shift workers who like such solitary activities as fishing, gardening, restoring old cars, or do-it-yourself home repair projects can find themselves happily suited for shift work.

Studies have also revealed that many shift workers turn these daytime hobbies into part-time jobs, hiring out their skills in gardening or home repair for extra money.

How Shift Work Affects Your Domestic Life

If you are married or living with someone, be assured that frequent shift work will severely test your relationship. If you and your mate have a solid, happy relationship, then you may well be able to cope with the added strain. But if your relationship is fragile, then the added stress of shift work could damage it beyond repair. Shift work not only creates new problems, it also magnifies existing ones. If you are in a tenuous relationship that you want to save, think twice about working the night beat.

The underlying cause of the strain that comes from shift work is a lack of synchronization between you and your partner. As we mentioned earlier, unless the two of you are shift workers on the same schedule, the effect of your differing hours is the same as living in different time zones. Finding the time to be together, whether it be for making small talk or making love, can be extremely difficult. The night shift also rules out attending most evening social functions together, which can cause resentment on both sides—from the shift worker who must forego the event and from the spouse who has to go to the function alone.

As we noted earlier, most research in the area of shift work has concentrated on male shift workers. Thus, most of the research on how shift work affects the *spouses* of shift workers has focused on women. However, studies indicate that no matter who is doing the shift work and who is staying home, both members of the couple must cope with a tremendous amount of stress.

Women who are married to shift workers report that their lives are greatly disrupted by their husbands' schedules. To keep their households functioning, they, like their husbands, must live in both the day and night worlds. Their most frequent complaints are: (1) an inability to do everyday work around the house because the husband is sleeping, (2) difficulty in keeping the children quiet while the husband sleeps, (3) fear and loneliness at night, (4) having to stay home in the evening, (5) problems scheduling meals, and (6) having to put up with a husband who is often cranky because he works shifts.

Complaints from the shift workers center less on specific problems and more on how they see themselves as family members. Men working shifts often worry that, because of their hours, they cannot adequately fulfill their traditional role as leader of the family. They worry that they are not home at night to protect their wives and children. They complain that they cannot be a proper companion to their wives, and they fear that their absence from home will result in undisciplined children who do not have a proper respect for authority.

In fact, some research does suggest that the children of shift workers do not do as well in school as other children. However, a major long-term study of 16,000 English schoolchildren indicates that the children of shift workers do just as well in school and are as emotionally well-adjusted as their peers. But the fear that their children will somehow turn out wrong persists among many shift workers.

While shift work creates many problems for a couple, it also can offer a few advantages. For example, Ted and Barbara, a couple in their late twenties with two young children, find they must both work to make ends meet. Ted works a regular day job as a building supervisor; Barbara works the 3:00 P.M. to midnight shift at a nursing home.

While working different shifts has strained their relationship from time to time, it has also meant that one of them is always home to take care of the children, except for two hours in the late afternoon, when the children go to a neighbor's home. Ted and Barbara have therefore avoided most of the day care worries and problems that plague other working parents. Also, they do not have to spend the money on day care that other couples do.

A night shift schedule also often means that a couple can be together during the day, which many find enjoyable. In addition, studies have found that men who are at home during the day tend to be more willing to share in the domestic chores traditionally done by their spouses—a decided advantage for the wives of shift-working husbands.

How Shift Work Affects Your Sleep

Sleep, or the lack of it, is the major problem for almost all shift workers. Shift workers get an average of seven fewer hours of sleep a week than their daytime colleagues. Trying to sleep in a room that is not completely dark, being awakened by noise from the daytime world, participating in day-oriented social or domestic events, holding a second daytime job—these and other experiences can cut into a shift worker's sleep time.

On a more basic level, shift work makes sleep difficult because it disrupts the body's natural circadian rhythms, including its sleep/wake cycle. With rhythms almost always out of phase and trying to readjust to the conflicting zeitgebers caused by shift work, the sleep/wake cycle is unable to settle into a routine pattern.

Even if you stay on the night shift for years and keep a fairly regular daytime sleeping schedule, you will still encounter problems. Indeed, as you get older, nature works against you and tends to make it even more difficult to sleep during the day—despite the fact that you need less sleep as you age (see chapter 3). The reason? As you get older you also tend to sleep less soundly and are more easily awakened—a serious problem for shift workers who must try to sleep during the morning hours. Various studies show that between 60 and 80 percent of shift workers complain that noise is their biggest obstacle to getting a good day's sleep.

For many shift workers, this lack of sleep is chronic, and they go through life with a sleep debt. They are always at least a little bit fatigued. In an effort to catch up on their sleep, shift workers often try to sleep longer than other people at weekends, but studies indicate that a few extra hours in bed on Saturday or Sunday morning does little to relieve the overall feeling of fatigue.

The chronic tiredness experienced by shift workers often has a negative effect on their work—particularly when the work they are doing is boring and repetitive. As anyone who has sat through a dull lecture knows, it is difficult, when fatigued, to stay awake when the activity you are involved in does not hold your interest.

Although you cannot completely avoid being tired if you work shifts regularly, you can help yourself cut down on sleep loss by going to bed as soon as possible after your shift. When you arrive home at 6:00 A.M. or 7:00 A.M., don't stay up for an hour or two reading the morning newspaper. Get right to sleep. It will give you some extra sleep time.

How Shift Work Affects Your Health

Shift work is bad for your health. It disrupts your biological rhythms, including your sleep cycle, and it causes a tremendous amount of emotional stress—all of which affects your physical health. Working odd hours also makes it difficult to eat decently, which only adds to the stress on your body. To make matters worse, you become more and more susceptible to the health problems associated with shift work as you age. So the longer you do it, the worse it gets.

Scientists view the impact of shift work on health from several perspectives, the most general of which concerns the overall impact of shift work, how it simply wears you down both physically and mentally and makes you more susceptible to illnesses. This general "lowering of well-being," as one researcher describes it, is related directly to poor eating habits, as well as to a lack of sleep.

All too often, shift work means eating bad food at irregular times. Shift workers often work when company cafeterias are closed, and although cafeteria food is often lacking in good nutrition, it is at least better than the junk food available in most vending machines. Shift workers, particularly those who work very late hours, tend to get much of their food from vending machines. If you are a shift worker and this sounds all too familiar, then the next time you look at a shriveled, chemically treated, plastic-wrapped hamburger through the little door of a vending machine, remember the old saying, "You are what you eat."

EASING THE STRAIN OF SHIFT WORK

Although you cannot totally avoid the emotional and physical problems that accompany shift work, there are several things you can do to help ease the strain.

• *Stay in shape.* One way to help your body cope better with the stress and strain of shift work is to get in shape through exercise. Shift work is not an athletic competition, but it does demand a certain amount of physical endurance. The better physical shape you are in, the better you can tolerate the disruption to your rhythms.

• *Eat well.* Pay close attention to what—and when—you eat. Instead of snacking your way through the night shift, try to have a regular "lunch" period. Also, make the effort to eat well. Instead of relying on vending machines, bring nutritious food from home to sustain you through your shift. Keep in mind that fatty foods and high-carbohydrate foods can make you sleepy; pack your lunch box with low-fat, high-protein foods instead. (For a list of healthful protein-rich foods and a discussion of why they help keep you alert, see chapter 7.) Also, watch your caffeine intake. A cup or two of coffee at the beginning of your shift may be helpful in keeping you alert through the night. However, avoid all caffeine during the last five hours of your shift; otherwise, you may find it difficult to fall asleep when you get home.

• *Take naps.* Sleeping on the job can get most workers fired, but in some Japanese factories, night shift workers are given a sleeping allowance that permits them to take short naps to refresh themselves. Their employers believe that allowing workers to take short naps is better than forcing them to struggle through the night trying to stay awake. Indeed, some scientists have recommended on-the-job sleeping periods for shift workers in industries involving public safety. One NASA researcher has even proposed that nap periods be established for individual members of airline crews on long overnight trips to avoid the danger of all members of a plane's crew falling asleep at the same time—something, unfortunately, that has happened. The message is clear: Naps help night shift workers. So if there is a safe and practical way you can doze for a few minutes on your shift, do it.

• *Guard your sleep.* Even under ideal conditions, it is difficult to get a long, deep sleep during the daytime because you are working against your natural rhythms. Make sure you have a dark, quiet room in which to sleep, and then set up a strict time to go to bed and stick to it. A lack of sleep is your worst enemy as a shift worker.

• *Get regular health checkups.* It is important to closely monitor your health if you are working shifts. A German study group composed of scientists and occupational health experts has developed the following recommendations for shift workers: If you will regularly be working a shift that extends to at least 3:00 A.M., then you should undergo a full physical examination with your physician *before* you start the shift. After six months on the job, check in again with your physician, and then have regular exams throughout your career. Workers between the ages of twenty-five and fifty should have checkups every five years; those between fifty and sixty, every two or three years; those older than sixty, every year.

The Decline and Fall of Your Health

People who become permanent night shift workers go through four distinct and predictable phases. Each phase has its impact on a worker's health. Here's what researchers say about these phases of declining health.

Adaptation phase (years 1–5). When you begin shift work, you face the emotional and physical strain of adjusting to the new lifestyle. Your body must alter its rhythms to try to get in sync with your new hours, you must adjust to getting less sleep, and the structure of your domestic and social life has to be rebuilt. The strain of these changes varies from person to person, but for the first five years, it is this overall stress that leads to most illnesses related to shift work.

Sensitization phase (years 5–20). No longer a rookie in your field of work, you are probably concerned with career advancement and your working conditions—added stressors on top of the ongoing ones associated with shift work. If you are married, it is likely you have children entering school, and you may be worrying about buying a home. It is during these years that many shift workers suddenly find they no longer like shift work. Because job satisfaction, or the lack of it, plays an important role in a worker's health, this dissatisfaction with shift work can accelerate the health problems associated with it.

Accumulation phase (years 20–40). At this time in your life, family, social, and financial problems are usually at a minimum, and life is on an even keel. Still, years of not getting enough sleep and eating poorly, as well as other stress-related problems, can begin to take their toll. Also, for reasons not entirely clear to researchers, it is during this period that workers, especially those in hazardous professions, begin to take more on-the-job risks. The accumulation of these factors can cause more work-related accidents and health problems.

Manifestation phase (years 40 +). This is the time when diseases that are most identified with shift work—chronic gastritis, gastric and duodenal ulcers, and other gastrointestinal disorders—reach their peak. In addition, other stress-related medical conditions that have been developing slowly during the accumulation phase, such as high blood pressure, continue to worsen under the strain of continued shift work.

Warning: Don't Work the Night Shift If You Have One of These Illnesses

Working an abnormal shift, especially the highly disruptive night shift, creates special problems for people who suffer from several specific illnesses. For the reasons discussed here, if you suffer from digestive disorders, diabetes, epilepsy, or a serious mental illness, you should absolutely avoid any type of regular shift work.

Chronic digestive tract disorders. Irregular mealtimes and a poor diet are two of the most notorious problems of shift work. Those two factors should serve as warnings to anyone who already suffers from ulcers, gastritis, or other digestive problems. Researchers also believe that the emotional stress many shift workers experience (due to domestic and social problems) contributes to gastric problems, as well. Even people with strong constitutions tend to go running for the antacids once they start shift work; if you are already predisposed to such problems, working the night shift may lead to serious complications.

Diabetes. People who suffer from diabetes must maintain a strictly controlled diet, both in terms of the timing and the content of their meals. That, for the many reasons we've already discussed, is nearly impossible while working shifts.

Epilepsy. Epileptic seizures and sleep are closely related (see chapter 6). Shift work almost guarantees a lack of sleep, which means an epileptic shift worker will experience an increase in seizures.

Severe mental illness. The lack of proper sleep that comes with shift work is also dangerous to people suffering from a variety of serious mental illnesses. Disturbed sleep patterns can trigger or worsen episodes of severe depression and other forms of mental illness (see chapter 4).

HOW SHIFT WORK AFFECTS YOUR WORK PERFORMANCE

For decades, factory owners, shift workers, and researchers have known that people who work the night shift are not as efficient as their daytime colleagues. Work is sloppier on the night shift; workers are more prone to have accidents and, in general, less gets done.

It is a problem that, for a long time, was left to researchers concerned with human performance and plant managers worried about production levels. That began to change at 4:00 A.M. on a Wednesday morning in the late winter of 1979. Tired workers halfway through their night shift made a series of unusual mistakes, and the result was an unprecedented release of radiation from the Three Mile Island nuclear power plant near Middletown, Pennsylvania. Suddenly, the debate over shift work and performance took on a more serious tone. What had been a problem for manufacturers trying to increase productivity was now a public safety issue.

While the Three Mile Island incident raised concerns over the impact of shift work on performance, it didn't result in any widespread changes in the way shift schedules are handled. Nor did the problem of tired workers making mistakes in critical jobs go away. Eight years later, at the Chernobyl nuclear power plant in the Soviet Union, halfway around the world from Three Mile Island, another series of human errors during an early morning experiment triggered a disastrous fire and the release of radiation from a nuclear reactor. This time, thirty-one lives were lost, tens of thousands of people were exposed to dangerous levels of radiation, and entire towns were made uninhabitable.

That these incidents happened, and that they were both caused in large part by human error, shouldn't be surprising. Nor should it be surprising if such life-threatening accidents happen again, given the growing use of shift work in increasingly complex jobs. For, as we have emphasized throughout this chapter, fatigue, sleep loss, and emotional stress are unavoidable for shift workers. And that translates into poor performance.

Laboratory studies of performance and shift work support these conclusions, and so do the relatively few field studies of workers on the job. The latter studies, six of them in all, ranging from 1949 to 1978, studied how fast operators answered a telephone switchboard, how many errors were made in reading meters, how often drivers nodded off while on the road, how quickly workers could join threaded parts together, how often train engineers missed warning signals, and how many minor accidents occurred in a hospital. In each of these studies, the night shift workers performed dramatically worse than their morning and afternoon shift colleagues.

The number of errors workers made in reading meters, for instance, increased substantially around midnight, then continued to rise until about 4:00 A.M. The error level then declined fairly rapidly until the shift ended four hours later. The other studies showed similar patterns.

Trying to determine why your performance is better at one time of the night than at another is not as simple as studying performance rhythms during the day. Shift workers, remember, suffer from a chronic lack of sleep and other stressors that do not usually afflict day workers. Until recently, researchers believed that performance rhythms were determined by the daily body temperature rhythm. They thought that, as your temperature rose, so did your ability to work error-free. They also believed that if you could get your temperature rhythm to climb during the night rather than during the day, your job performance would rise to the same level as that of a day worker.

Today, however, most chronobiologists believe that, although the two rhythms may travel on parallel paths, the temperature and performance rhythms are not linked, but instead function independently of one another. Indeed, the temperature rhythm of a worker who switches from a daytime job to the night shift will take about twelve days to get back into sync, but the worker's ability to perform complex, cognitive tasks will adjust within a few days. On the other hand, the worker's ability to do simpler tasks, such as putting a screw in a machine or watching for defective parts on an assembly line, will take even longer than the temperature rhythm to adjust.

Many workers, therefore, find that their ability to figure out complicated problems is actually quite good during their first few "days" on the night shift. Within a few days, however, that extra ability seems to fade away, and errors and fatigue set in.

NOT ALL SHIFT SCHEDULES ARE ALIKE

Employers have devised many types of shift schedules for their workers over the years. The three most common ones are the permanent shift, the slowly rotating shift, and the rapidly rotating shift.

Workers on the *permanent shift* stick to that shift week after week and do not rotate off it. People who work a *slowly rotating shift* change the hours of their shift weekly: One week they work the day shift; the next week, the evening shift; and the third week, the night shift. Then it's back to the day shift again. Workers assigned a *rapidly rotating shift* make much quicker shift changes. They spend only two days on each shift (which adds up to a six-day work week), followed by two days off.

The slowly rotating shift is the most stressful of the three types of shifts because it is the most confusing one for your rhythms. Just when your rhythms are beginning to commit to a set of new zeitgebers, you switch to another shift and throw them into disarray again. It has the same effect on your body as flying across seven or eight time zones every five days. The constant change gives you a chronic case of shift lag. Your body never has a chance to catch up to your changing hours, and your performance will be consistently low.

The other two types of shifts offer both advantages and disadvantages as far as performance is concerned. It depends on what type of work is involved. If your work involves complex mental tasks, such as computer programming or other kinds of problem solving, you are much better working a rapidly rotating shift. Most people on these quick two-day shifts tend to keep themselves oriented to the daytime. As a result, their biological rhythms also stay committed to a day schedule. Not only is that less stressful on the body, but it also means that people who work a rapidly rotating shift will benefit from the mysterious intellectual boost that occurs for a few nights after moving quickly to the night shift.

For more simple, repetitious types of work that require physical dexterity, such as assembly line work, the permanent, or long-term, shift seems best. Physical dexterity takes as long as two weeks to adjust to a new schedule. Although your job performance will never be as good at night as it would be during the day, working a permanent night shift at least gives your rhythms the time to adjust as much as possible to the new schedule.

Tips for Coping with Shift Work

Because shift work can have such a detrimental effect on all aspects of your life, the best tip for coping with shift work is not to do it in the first place! However, if you must work a shift, either permanently or temporarily, here is a summary of what you can do to lesson its impact.

• If you have a choice, choose a rapidly rotating shift over a slowly rotating one.

• Commit yourself to your shift. As much as possible, alter all other aspects of your life to fit your new schedule.

• Go to sleep as soon as you get home from an evening or night shift.

• Stick to a nutritious diet. Avoid relying on vending machines for your night "lunch."

• Get plenty of exercise. Your body will need all the strength it can get.

• Be alert to the fact that you will be more prone to accidents and other errors while working the night shift.

Ten

The Beat of the Future

For though the outward man perish,
yet the inward man is renewed day by day.
—II CORINTHIANS 4:16

Our ancient ancestors followed the beat of their inner rhythms. They ate when they were hungry, slept when they became tired, and labored and rested according to the urgings of their bodies. The sun and the seasons were their only "clocks," telling them when to hunt animals, when to plant crops, and when to start building shelters for winter. Like the animals they shared their habitats with, they were in synchronization with the natural world.

In the highly technological world in which we live today, most of us ignore the rhythms within us and tend to discount the external ones, as well. We eat at set times during the day, whether we are particularly hungry or not; we work at all hours of the day and night; and we often defy our body's desire to sleep by staying up into the small hours of the morning. Electricity, central heating, air conditioning, jet travel, and a host of other technological by-products, including such things as all-night restaurants, stores, and nightclubs, have encouraged us to lead lives that are out of sync with nature's rhythms.

Yet, as we have seen, we need to get back in touch with our biological rhythms—and with those of nature—for the sake of our physical health and mental well-being. That doesn't mean returning to the primitive lifestyle of our ancestors, but it does mean taking a new look at the way our lives are structured and making readjustments in our schedules that take into account our bodies' rhythmic fluctuations. It also means setting a daily pace to our lives that is more in step with the natural pacemakers within us.

Institutions, as well as individuals, need to be aware of the impact internal rhythms have on our lives. More companies need to instigate flex-time policies that allow employees to start work at staggered times—say, between 7:00 A.M. and 10:00 A.M.—as long as they put in an eight-hour day once they arrive. Both morning people and evening people could then schedule their workday according to their inner rhythms. In addition, more companies should keep circadian rhythms in mind when planning their shift work schedules. Police officers in Philadelphia, for example, recently instigated a circadian-conscious shift rotation with overwhelmingly positive results. The officers report being less sleepy and less prone to making errors.

Chronobiology is also beginning to make inroads into our medical institutions, where it may have its biggest impact on our lives. As we saw in Chapter 6, some hospitals are already beginning to experiment with the timing of cancer drugs, and some psychiatrists are taking body rhythms into consideration when treating certain forms of depression. In the future, these kinds of chronotherapies may become commonplace.

Widespread awareness of the importance of chronobiology, however, appears to be a long way off. Airline pilots and junior doctors, for example, still work long and rhythm-disruptive shifts, leaving them more susceptible to making life-threatening errors. Important summits between nations are conducted with only a perfunctory concern about jet lag. And the research labs of major pharmaceutical companies continue to test drugs with no adjustments for circadian rhythms.

In the future, as knowledge about chronobiology grows, the practices of our major institutions will change and biological rhythms will be taken into account. In the meantime, each of us can increase our knowledge about our individual rhythms. Chart your temperature, blood pressure, and other daily cycles. If you are a woman, trace the changes that occur to your body monthly. Learn how to listen to the inner beats of your body; let them set the pace of your day. You will live a healthier—and happier—life.

Bibliography

General

CAMPBELL, JEREMY. *Winston Churchill's Nap*. New York: Simon and Schuster, 1986.

"Chronobiology: A Science in Tune with the Rhythms of Life." Minneapolis: Earl Bakken, 1986.

CONROY, R. T. W. L., and J. N. MILLS. *Human Circadian Rhythms*. London: J. & A. Churchill, 1970.

LUCE, GAY. *Biological Rhythms in Psychiatry and Medicine*. Washington, DC: National Institute of Mental Health, US Department of Health, Education and Welfare, 1970.

MONK, TIMOTHY. "Research Methods of Chronobiology." In *Biological Rhythms, Sleep and Performance*. Chichester, England: John Wiley & Sons, 1982.

PALMER, JOHN. *Introduction to Biological Rhythms*. New York: Academic Press, Inc., 1976.

PAULY, JOHN. "Chronobiology: Anatomy in Time." *American Journal of Anatomy*, 1983, Vol. 168, pp. 365–388.

RUSAK, BENJAMIN, AND IRVING ZUCKER. "Biological Rhythms and Animal Behavior." *Annual Review of Psychology*. Vol. 26, 1975, pp. 137–171.

SCHMECK, HAROLD. "Studying Life's Rhythms, Scientists Find Surprises." *New York Times*, April 15, 1986.

WESTON, LEE. *Body Rhythm*. New York and London: Harcourt Brace Jovanovich, 1979.

WINFREE, ARTHUR. *The Timing of Biological Clocks*. New York: *Scientific American* Library, 1987.

Chapter 1: The Times of Your Life

ASCHOFF, J., M. FATRANSKA, AND H. GIEDKE. "Human Circadian Rhythms in Continuous Darkness: Entrainment by Social Cues." *Science,* January 15, 1971, pp. 213–215.

BROWN, FRANK A. "The 'Clocks' Timing Biological Rhythms." *American Scientist,* November/December 1972, pp. 756–765.

FOLKARD, SIMON. "Circadian Rhythms and Human Memory." In *Rhythmic Aspects of Behavior.* Hillsdale, New Jersey: Lawrence Erlbaum Associates, Inc., 1982.

GARDNER, RALPH. "The Mania of the Full Moon." *Cosmopolitan,* June 1983, pp. 202–206.

HALBERG, FRANZ, EUGENE JOHNSON, WALTER NELSON, WALTER RUNGE, AND ROBERT SOTHERN. "Autorhythmometry: Procedures for Physiologic Self-Measurements and Their Analysis." *Physiology Teacher,* January 1972, Vol. I.

KOLATA, GINA. "Genes and Biological Clocks." *Science,* December 1985, pp. 1151–1152.

PALMER, JOHN. "Biological Clocks of the Tidal Zone." *Scientific American,* February 1975, pp. 70–79.

REINBERG, ALAIN, AND MICHAEL SMOLENSKY. "Introduction to Chronobiology." In *Biological Rhythms and Medicine.* New York: Springer-Verlag, 1983.

RESTAK, RICHARD. "Master Clock of the Brain and Body." *Science Digest,* November 1984.

SAUNDERS, D. S. "The Biological Clock of Insects." *Scientific American,* February 1976, pp. 114–121.

SHEPHARD, ROY. "Sleep, Biorhythms and Human Performance." *Sports Medicine,* 1984, Vol. 1, pp. 11–37.

SWEENEY, BEATRICE. "Biological Clocks—An Introduction." *BioScience,* July/August 1983, pp. 424–425.

Chapter 2: Your Daily Ups and Downs

FOLKARD, SIMON, AND TIMOTHY MONK. "Circadian Rhythms in Human Memory." *British Journal of Psychology,* 1980, Vol. 71, pp. 295–307.

HALES, DIANNE. "Sunrise, Sunset: The Rhythms of Your Life." In *The Complete Book of Sleep.* Reading, Massachusetts: Addison-Wesley Publishing Co., 1981.

MONK, TIMOTHY, VICTORIA LENG, SIMON FOLKARD, AND ELLIOT WEITZMAN. "Circadian Rhythms in Subjective Alertness and Core

Body Temperature." *Chronobiologia,* January–March 1983, Vol. X, pp. 49–55.

MONK, TIMOTHY, ELLIOT WEITZMAN, JEFFREY FOOKSON, MARGARET MOLINE, RICHARD KRONAUER, AND PHILIPPA GANDER. "Task Variables Determine Which Biological Clock Controls Circadian Rhythms in Human Performance." *Nature,* August 11, 1983, Vol. 304, pp. 543–545.

MONK, TIMOTHY, JEFFREY FOOKSON, MARGARET MOLINE, AND CHARLES POLLAK. "Diurnal Variation in Mood and Performance in a Time-Isolated Environment." *Chronobiology International,* 1985, Vol. 2, pp. 185–193.

MONK, TIMOTHY, ELLIOT WEITZMAN, JEFFREY FOOKSON, AND MARGARET MOLINE. "Circadian Rhythms in Human Performance Efficiency Under Free-Running Conditions." *Chronobiologia,* October–December 1984, Vol. XI, pp. 343–354.

MONK, TIMOTHY, AND SIMON FOLKARD. "Concealed Inefficiency of Late-Night Study." *Nature,* May 25, 1978, Vol. 273, pp. 296–297.

MONK, TIMOTHY, JEFFREY FOOKSON, JACOB KREAM, MARGARET MOLINE, CHARLES POLLAK, AND MURIEL WEITZMAN. "Circadian Factors During Sustained Performance: Background and Methodology." *Behavior Research Methods, Instruments, & Computers,* 1985, pp. 19–26.

MONK, TIMOTHY. "Circadian Rhythms in Human Performance." *Directions in Psychiatry,* Vol. 4, Lesson 19. New York: Hatherleigh Company, Ltd., 1984.

MONK, TIMOTHY, AND VICTORIA LENG. "Interactions Between Inter-Individual and Inter-Task Differences in the Diurnal Variation of Human Performance." *Chronobiology International,* 1986, Vol. 3, pp. 171–177.

MONK, TIMOTHY, JEFFREY FOOKSON, MARGARET MOLINE, AND CHARLES POLLAK. "Diurnal Variation in Mood and Performance in a Time-Isolated Environment." *Chronobiology International,* 1985, Vol. 2, pp. 185–193.

WINGET, CHARLES, CHARLES DeROSHIA, AND DANIEL HOLLEY. "Circadian Rhythms and Athletic Performance." *Medicine and Science in Sports and Exercise,* 1985, Vol. 17, pp. 494–516.

WURTMAN, RICHARD. "The Effects of Light on the Human Body." *Scientific American,* July 1975, pp. 69–77.

Chapter 3: The Importance of Sleep

DEMENT, WILLIAM. *Some Must Watch While Some Must Sleep.* San Francisco: San Francisco Book Co., Inc., 1976.

"Dreams and Memories." *Science News,* July 3, 1982, Vol. 122, p. 10.

FORD, BARBARA. "Making Sense of Sleep." *National Wildlife,* August/ September 1985.

FORD, NORMAN. *Good Night.* Gloucester, Massachusetts: Para Research, Inc., 1983.

HALES, DIANNE. *The Complete Book of Sleep.* Reading, Massachusetts: Addison-Wesley Publishing Co., 1981.

HAURI, PETER. "The Sleep Disorders." Kalamazoo, Michigan: The Upjohn Company, 1977.

HICKS, ROBERT A. "Normal Insomnia: Its Benefits and Its Costs." Annual Scholar's Address of the Western Psychological Association's annual convention, 1983, hosted by San Jose State University.

HOLZMAN, DAVID. "Sleeping to Help Fight Disease." *Insight,* December 15, 1986, pp. 58–59.

LAMBERG, LYNNE. *The American Medical Association Guide to Better Sleep.* New York: Random House, 1984.

LUCE, GAY GAER, AND JULIUS SEGAL. *Sleep.* New York: Coward-Mc-Cann, Inc., 1966.

"The Nightmare of Sleepless Nights." *Science News,* November 1, 1986, Vol. 130, p. 280.

ROVNER, SANDY. "It's Possible That a Bad Night's Sleep Will Work Wonders." *Minneapolis Star & Tribune,* June 27, 1985.

STEINHART, PETER. "Sleeping on the Job." *National Wildlife,* October/ November 1982.

Chapter 4: The Measure of Your Moods

ALBIN, ROCHELLE SEMMEL. "The Holiday Blues—A Christmas Fable?" *Psychology Today,* December 1981.

ALPER, JOSEPH. "Biology and Mental Illness." *Atlantic Monthly,* December 1983.

BOWER, B. "Winter Depression: Rise and Shine?" *Science News,* December 20 and 27, 1986, Vol. 130, p. 390.

CORFMAN, EUNICE. "Depression, Manic-Depressive Illness, and Biological Rhythms." Report published by the National Institute of Mental Health, 1979.

"Deflating Christmyth Depression." *Science News,* January 9, 1982, Vol. 121, p. 24.

"Depression: Brain Chemistry Gone Awry." *Science Digest,* December 1982.

DULLEA, GEORGIA. "Shedding Light on Dark-Day Blues." *New York Times,* December 19, 1985.

GREENBERG, JOEL. "Cracking the Cycles of Depression and Mania." *Science News,* November 25, 1978, Vol. 114, p. 367.

HERBERT, W. "Throwing Light on the Winter Blues." *Science News,* March 27, 1982, Vol. 121, p. 212.

JENNER, F. A. "Periodic Psychoses in the Light of Biological Rhythm Research." *International Review of Neurobiology,* 1968, Vol. 11, pp. 129–169.

KLINE, NATHAN, ARISTIDE ESSER, PER VESTERGAARD, AND FRANZ HALBERG. "Techniques for Assessing Biological Rhythms in Psychiatry and Medicine." In *Mental Health Program Reports No. 3.* National Institutes of Mental Health, January 1969.

KRIPKE, DANIEL, DANIEL MULLANEY, MARTHA ATKINSON, AND SANFORD WOLF. "Circadian Rhythm Disorders in Manic-Depressives." *Biological Psychiatry,* 1978, Vol. 13, pp. 335–351.

LEWY, ALFRED, AND ROBERT SACK. "Minireview: Light Therapy and Psychiatry." *Proceedings of the Society for Experimental Biology and Medicine,* 1986, Vol. 183, pp. 11–18.

LEWY, ALFRED, ROBERT SACK, STEPHEN MILLER, AND TANA HOBAN. "Antidepressant and Circadian Phase-Shifting Effects of Light. *Science,* January 16, 1987, Vol. 235, pp. 352–354.

MAUGH, THOMAS. "Biochemical Markers Identify Mental States." *Science,* October 1981, Vol. 214, pp. 39–41.

OLDENBURG, DON. "Can the Sun Cure Ills Instead of Causing Them?" *Minneapolis Star & Tribune,* May 12, 1986.

PEKKANEN, JOHN. "Beating the Winter Blues: A New Light Can Bring What Seems Like Sunshine to Those Who Get Depressed by the Short Days of December." *The Washingtonian,* December 1983.

RICHTER, CURT. "The Role of Biological Clocks in Mental and Physical Health." In *Mental Health Program Reports No. 3.* National Institute of Mental Health, January 1969.

ROCKWELL, D. A., C. M. WINGET, L. S. ROSENBLATT, E. A. HIGGINS, AND N. W. HETHERINGTON. "Biological Aspects of Suicide: Circadian Disorganization." *Journal of Nervous and Mental Disease.* 1978, Vol. 166, pp. 851–858.

SCHMECK, HAROLD. "Manic-Depressive Cycle Tied to 'Clock' Defect." *New York Times,* December 5, 1978.

"Serotonin: A Natural Anti-Depressant?" *Science News,* April 10, 1982, Vol. 121, p. 251.

STAMPS, DAVID. "Give Me a Light. A Minnesota Light." *Minneapolis Star & Tribune Sunday Magazine,* December 22, 1985.

TALAN, JAMIE. "Sleeping Habits May Have Effect on Depression." *Minneapolis Star & Tribune,* June 12, 1986.

WEHR, THOMAS. "Circadian Rhythm Disturbances in Depression and Mania." In *Rhythmic Aspects of Behavior*. Hillsdale, New Jersey: Lawrence Erlbaum Associates, 1982.

WEHR, THOMAS, ANNA WIRZ-JUSTICE, AND FREDERICK GOODWIN. "Phase Advance of the Circadian Sleep-Wake Cycle as an Antidepressant." *Science,* November 1979, Vol. 206, pp. 710–713.

WEITZMAN, ELLIOT. "Biological Rhythms: Indices of Pain, Adrenal Hormones, Sleep, and Sleep Reversal." In *Mental Health Program Reports No. 3.* National Institutes of Mental Health, January 1969.

Chapter 5: The Rhythms of Reproduction

ASSO, DOREEN. *The Real Menstrual Cycle.* Chichester, England: John Wiley and Sons, 1983.

Behavior and the Menstrual Cycle. Edited by Richard C. Friedman. New York: Marcel Dekker, Inc., 1982.

BONEN, AREND, AND HANS KEIZER. "Athletic Menstrual Cycle Irregularity: Endocrine Response to Exercise and Training." *The Physician and Sportsmedicine,* August 1984, pp. 78–90.

BRODY, JANE. "Diet, Exercise, Stress Management Best PMS Remedies." *Minneapolis Star & Tribune,* April 27, 1986.

BROOKES-GUNN, J., JANINE GARGIULO, AND MICHELL WARREN. "The Effect of Cycle Phase on the Performance of Adolescent Swimmers." *The Physician and Sportsmedicine,* March 1986, Vol. 14, pp. 182–192.

BUTLER, ELLEN. "The Amenorrheic Athlete." *The Melpomene Report,* June 1985.

CAVANAGH, JOHN. "Rhythm of Sexual Desire in Women." *Medical Aspects of Human Sexuality,* February 1969, Vol. III, pp. 29–39.

CHEN, A. "Marijuana and the Reproductive Cycle." *Science News,* March 26, 1983, Vol. 123, p. 197.

ESTON, ROGER. "The Regular Menstrual Cycle and Athletic Performance." *Sports Medicine,* 1984, Vol. 1, pp. 431–445.

FREEDMAN, SHANNA, SAVITRI RAMCHARAN, AND ELIZABETH HOAG. "Some Physiological and Biochemical Measurements Over the Menstrual Cycle." In *Biorhythms and Human Reproduction.* Chichester: John Wiley and Sons, 1974.

HARKNESS, R. A. "Variations in Testosterone Excretion by Man." In *Biorhythms and Human Reproduction,* Chichester, England: John Wiley and Sons, 1974.

HERSEY, REX. "Emotional Cycles in Man." *Journal of Mental Science,* 1931, Vol. LXXVII, pp. 151–169.

HUTT, S. J., G. FRANK, W. MYCHALKIW, AND M. HUGHES. "Perceptual-

Motor Performance During the Menstrual Cycle." *Hormones and Behavior,* 1980, Vol. 14, pp. 116–125.

JONES, JULIE. "PMS: Treatment." *The Melpomene Report,* Fall 1986.

JONES, RICHARD E. *Human Reproduction and Sexual Behavior.* Englewood Cliffs, New Jersey: Prentice-Hall, Inc., 1984.

KAISER, IRWIN. "Synchronizers of Reproductive Function in Men and Women." In *Biorhythms and Human Reproduction.* Chichester, England: John Wiley and Sons, 1974.

KIHLSTROM, J. E. "A Monthly Variation in Beard Growth in One Man." *Life Sciences,* March 22, 1971, Vol. 10, pp. 321–324.

MCCLINTOCK, MARTHA. "Menstrual Synchrony and Suppression." *Nature,* January 22, 1971, Vol. 229, pp. 244–245.

PARLEE, MARY BROWN. "New Findings: Menstrual Cycles and Behavior." *Ms.,* September 1982.

RAMEY, ESTELLE. "Men's Cycle: They Have Them Too, You Know." *Ms.,* Spring 1972.

RAU, JULIE. "Athletic Amenorrhea: A Review of Literature." Report issued by the Melpomene Institute, May 1983.

REINISCH, JUNE. " 'Menstrual Synchrony' Documented Fact: The Kinsey Report." *Minneapolis Tribune,* April 15, 1984.

SMOLENSKY, MICHAEL, ALAIN REINBERG, RUFUS LEE, AND JOHN MCGOVERN. "Secondary Rhythms Related to Hormonal Changes in the Menstrual Cycle: Special Reference to Allergology." In *Biorhythms and Human Reproduction.* Chichester, England: John Wiley and Sons, 1974.

SOJOURNER, MARY. "A Survivor's Guide to Menstrual Cycle Changes/ PMS." Rochester, New York: Planned Parenthood of Rochester and Monroe County, Inc., 1983.

SOLOMON, SUSAN, MINDY KURZER, AND DORIS HOWES CALLOWAY. "Menstrual Cycle and Basal Metabolic Rate in Women." *American Journal of Clinical Nutrition,* Vol. 36, October 1982, pp. 611–616.

SOMMER, BARBARA. "How Does Menstruation Affect Cognitive Competence and Psychophysiological Response?" In *Lifting the Curse of Menstruation: A Feminist Appraisal of the Influence of Menstruation on Women's Lives.* New York: Haworth Press, 1983.

SOUTHAN, ANNA, AND FLORANTE GONZAGA. "Systemic Changes During the Menstrual Cycle." *American Journal of Obstetrics and Gynecology,* January–April 1965, Vol. 91, pp. 142–165.

UDRY, J. RICHARD. "Distribution of Coitus in the Menstrual Cycle." *Nature,* November 9, 1968, Vol. 220, pp. 593–596.

VERMEULEN, A., L. VERDONCK, AND F. COMHAIRE. "Rhythms of the Male Hypothalamo-Pituitary-Testicular Axis." In *Biorhythms and Hu-*

man Reproduction. Chichester, England: John Wiley and Sons, 1974.

WEIDEGER, PAULA. *Menstruation and Menopause*. New York: Alfred Knopf, 1980.

ZIMMERMAN, STANLEY, MAUREEN MAUDE, AND MARC MOLDAWER. "Frequent Ejaculation and Total Sperm Count, Motility, and Form in Humans." *Fertility and Sterility*, 1965, Vol. 16, pp. 342–345.

Chapter 6: You Are When You Eat

BRODY, JANE. "Survey Finds That Number of Americans Who Skip Breakfast Has Started to Drop." *Minneapolis Star & Tribune*, May 3, 1987.

FERNSTROM, JOHN, AND RICHARD WURTMAN. "Nutrition and the Brain." *Scientific American*, February 1974, Vol. 230, pp. 84–91.

GATTY, RONALD. *The Body Clock Diet*. New York: Simon and Schuster, 1978.

KAGAN, DANIEL. "Mind Nutrients." *Omni*, May 1985.

PEKKANEN, JOHN. "Researchers Are Discovering That Foods Can Alter Moods." *Minneapolis Star & Tribune*, July 2, 1986.

"Proposed Weight Loss Program Is Based on Timing of Meals." *Postgraduate Medicine*, March 1986, Vol. 79, p. 352.

"Researchers: Caffeine Helps and Hinders." *Minneapolis Star & Tribune*, March 11, 1987.

REINBERG, ALAIN. "Chronobiology and Nutrition." In *Biological Rhythms and Medicine*. New York: Springer-Verlag, 1983.

WURTMAN, JUDITH. *Managing Your Mind and Mood Through Food*. New York: Rawson Associates, 1987.

Chapter 7: Keeping a Healthy Beat

BRODY, ROBERT. "Drug-Taking Efficiency Goes Like Clockwork." *Minneapolis Star & Tribune*, October 6, 1986.

"Chronobiology and the Digestive System." Proceedings of a meeting held at the University of Minnesota in September 1981. NIH Publication No. 84-857, US Department of Health and Human Services, National Institutes of Health, May 1984.

COPE, LEWIS. "Three Cancers Linked to Seasonal Patterns." *Minneapolis Star & Tribune*, February 13, 1984.

HALBERG, FRANZ. "Implications of Biologic Rhythms for Clinical Practice." *Hospital Practice*, January 1977, Vol. 12, pp. 139–149.

HALBERG, JULIE, FRANZ HALBERG, AND CHARLES LEACH. "Variability of Human Blood Pressure with Reference Mostly to the Non-Chronobiologic Literature." *Chronobiologia*, July–September 1984, Vol. XI, pp. 205–216.

HRUSHESKY, WILLIAM. "Circadian Timing of Cancer Chemotherapy." *Science*, April 5, 1985, Vol. 228, pp. 73–75.

HRUSHESKY, WILLIAM. "The Clinical Application of Chronobiology to Oncology." *American Journal of Anatomy*, 1983, Vol. 168, pp. 519–542.

HRUSHESKY, WILLIAM. "Circadian Timing of Cancer Chemotherapy." *Science*, April 5, 1985, Vol. 228, pp. 73–75.

MOORE-EDE, MARTIN, CHARLES CZEISLER, AND GARY RICHARDSON. "Circadian Timekeeping in Health and Disease (Part I)." *New England Journal of Medicine*, August 25, 1983, Vol. 309, pp. 469–475.

MOORE-EDE, MARTIN, CHARLES CZEISLER, AND GARY RICHARDSON. "Circadian Timekeeping in Health and Disease (Part 2)." *New England Journal of Medicine*, September 1, 1983, Vol. 309, pp. 530–535.

MULLER, E., PETER STONE, ZOLTAN TURI, JOHN RUTHERFORD, CHARLES CZEISLER, CORETTE PARKER, W. KENNETH POOLE, EU-GENE PASSAMANI, ROBERT ROBERTS, THOMAS ROBERTSON, BUR-TON SOBEL, JAMES WILLERSON, EUGENE BRAUNWALD, AND THE MILIS STUDY GROUP. "Circadian Variation in the Frequency of Onset of Acute Myocardial Infarction." *New England Journal of Medicine*, November 21, 1985, Vol. 313, pp. 1315–1322.

PALMER, JOHN. "Human Rhythms." *BioScience*, February 1977, Vol. 27, pp. 93–99.

REINBERG, ALAIN. "Clinical Chronopharmacology: An Experimental Basis for Chronotherapy." In *Biological Rhythms and Medicine*. New York: Springer-Verlag, 1983.

SLOVUT, GORDON. "Anoka Man to Be First Cancer Patient Treated at Home with Drug Pump." *Minneapolis Star & Tribune*, September 30, 1986.

SLOVUT, GORDON. "Ovarian Cancer Patients Getting a Better Chance." *Minneapolis Tribune*, July 24, 1983.

SMOLENSKY, MICHAEL. "Aspects of Human Chronopathology." In *Biological Rhythms and Medicine*. New York: Springer-Verlag, 1983.

"Table Ronde, No. 54: Chronopharmacology and Chronotherapy." Papers presented at meeting of Tables Rondes, Roussel Uclaf; Institut Scientifique, Roussel, Paris, December 12–13. Published by Institut Scientifique, 1986.

"Time: The Fourth Dimension of Medicine" (Vols. 1 and 2). Reports from the Symposium on the Medical Implications of Chronobiology held in New Orleans, Louisiana, November 2–4, 1973. Published by The Upjohn Company, 1974.

Chapter 8: Rhythms Off Beat: Jet Lag

ANDERSON, IAN. "Pilots May Benefit from Sleeping at Work." *New Scientist,* February 20, 1986.

BLAKESLEE, ALTON. "Probing the Secrets of Jet Lag." *Science Digest,* November 1984.

BRODY, JANE. "Problems Presented by Jet Lag Can Put Damper on Vacation." *Minneapolis Star & Tribune,* August 18, 1985.

EHRET, CHARLES, AND LYNNE WALLER SCANLON. *Overcoming Jet Lag.* New York: Berkley Books, 1983.

FOUSHEE, H. CLAYTON. "Assessing Fatigue." *Air Line Pilot,* May 1986.

GANDER, PHILIPPA, GRETE MYHRE, R. CURTIS GRAEBER, HARALD ANDERSEN, AND JOHN LAUBER. "Crew Factors in Flight Operations: I. Effects of 9-Hour Time-Zone Changes on Fatigue and the Circadian Rhythms of Sleep/Wake and Core Temperature." NASA Technical Memorandum 88197, National Aeronautics and Space Administration, December 1985.

GRAEBER, R. CURTIS. "Alterations in Performance Following Rapid Transmeridian Flight." In *Rhythmic Aspects of Behavior.* Hillsdale, New Jersey: Lawrence Erlbaum Associates, 1982.

GRAEBER, R. CURTIS. "Crew Factors in Flight Operations: IV. Sleep and Wakefulness in International Aircrews." NASA Technical Memorandum 88231, National Aeronautics and Space Administration, February 1986.

GRAEBER, R. CURTIS, H. CLAYTON FOUSHEE, PHILIPPA GANDER, AND GERALD NOGA. "Circadian Rhythmicity and Fatigue in Flight Operations." *Journal of UOEH.* Published by the University of Occupational and Environmental Health, Japan, March 1, 1985, Volume 7 Supplement, pp. 122–129.

GRAEBER, R. CURTIS. "Jet Lag: The Circadian System in Transition." Speech presented at the APA Annual Meeting, Los Angeles, California, August 24, 1985.

HICKS, ROBERT, KRISTIN LINDSETH, AND JAMES HAWKINS. "Daylight Saving-Time Changes Increase Traffic Accidents." *Perceptual and Motor Skills,* 1983, Vol. 56, pp. 64–66.

MCFARLAND, ROSS. "Air Travel Across Time Zones." *American Scientist,* January/February 1975, Vol. 63, pp. 23–30.

MOHLER, STANLEY, J. ROBERT DILLE, AND H. L. GIBBONS. "The Time Zone and Circadian Rhythms in Relation to Aircraft Occupants Taking Long-Distance Flights." *American Journal of Public Health and the Nation's Health,* August 1968, Vol. 58, pp. 1404–1409.

PRESTON, F. S., S. C. BATEMEN, R. V. SHORT, AND R. T. WILKINSON.

"The Effects of Slying and of Time Changes on Menstrual Cycle Length and on Performance in Airline Stewardesses." In *Biorhythms and Human Reproduction*. Chichester, England: John Wiley and Sons, 1974.

SCHEVING, L. E., AND F. HALBERG (eds.). *Chronobiology: Principles and Applications to Shifts in Schedules*. Alphen aan den Rijn, the Netherlands and Rockville, Maryland: Sijthoff and Noordhoff, 1980.

"Setting the Clock." *Scientific American,* July 1986, p. 66.

SIEGEL, PETER, SIEGFRIED GERATHEWOHL, AND STANLEY MOHLER. "Time Zone Effects." *Science,* June 13, 1969, Vol. 164, pp. 1249–1255.

SNYDER, SCOTT. "Isolated Sleep Paralysis After Rapid Time Zone Change ('Jet Lag') Syndrome." *Chronobiologia,* October–December 1983, Vol. X, pp. 377–379.

ZARLEY, CRAIG. "The Daylight Savings Zone." *California Magazine,* May 1983.

Chapter 9: Rhythms Off Beat: Shift Work

FOLKARD, SIMON, AND TIMOTHY MONK. "Towards a Predictive Test of Adjustment to Shift Work." *Ergonomics,* 1979, Vol. 22, pp. 79–91.

FOLKARD, SIMON, TIMOTHY MONK, AND MARY LOBBAN. "Short and Long-term Adjustment of Circadian Rhythms in 'Permanent' Night Nurses." *Ergonomics,* 1978, Vol. 21, pp. 785–799.

FOLKARD, SIMON, AND TIMOTHY MONK (eds.). *Hours of Work: Temporal Factors in Work-Scheduling*. Chichester, England: John Wiley & Sons, 1985.

KNOX, RICHARD. "Light Adjusts Sleep Pattern, Study Shows." *Boston Globe,* August 1, 1986.

MONK, TIMOTHY. "Spring and Autumn Daylight Saving Time Changes: Studies of Adjustment in Sleep Timings, Mood, and Efficiency." *Ergonomics,* 1980, Vol. 23, pp. 167–178.

MONK, TIMOTHY. "Circadian Rhythms and Shiftwork." In *Stress and Fatigue in Human Performance*. Chichester: John Wiley & Sons, 1983.

MONK, TIMOTHY, AND DONALD TEPAS. "Shift Work." In *Job Stress and Blue Collar Work*. Chichester: John Wiley & Sons, 1985.

NAITOH, PAUL. "Chronobiologic Approach for Optimizing Human Performance." In *Rhythmic Aspects of Behavior*. Hillsdale, New Jersey: Lawrence Erlbaum Associates, 1982.

"Shift Work and Health . . . A Symposium." Papers presented at a symposium sponsored by the National Institute for Occupational

Safety and Health, June 12–13, 1975, in Cincinnati, Ohio. Published by the US Department of Health, Education and Welfare, July 1976.

TASTO, DONALD, AND MICHAEL COLLIGAN. *Shift Work Practices in the United States.* Report prepared for the National Institute for Occupational Safety and Health, Division of Biomedical and Behavioral Science, March 1977.

WOJTCZAK-JAROSZOWA, JADWIGA. *Physiological and Psychological Aspects of Night and Shift Work.* Published by US Department of Health, Education and Welfare, National Institute for Occupational Safety and Health, December 1977.

Index